Netherfield Road North
Everton Terrace
Netherfield Road
St Anthonys Church
St Martins Church
Great Homer St
Richmond Row
Christian Street
BYROM STREET
St Johns Lane
LIME STREET
ST GEORGE'S HALL
Railway Entrance
Copperas Hill
Ranelagh Pl
Hatton Garden
Tithebarn Street
DALE STREET
Whitechapel
HANOVER STREET
SAILORS HOME
Town Hall
WATER STREET
Chapel St
ST NICHOLAS CH.
GOREE
Brunswick St
WAREHOUSES
STRAND STREET
Canning Pl
NEW CUSTOM HOUSE
Canning Place
East Side Salthouse Dock
Dukes Place
Dock Police
CANNING DOCK
GEORGES DOCK
SALTHOUSE DOCK
Public Baths
GEORGES DOCK PARADE
MANCHESTER DOCK
ALBERT DOCK PARADE
Dukes Yard & PIER
KINGS
BIRKENHEAD HOTEL

TAKE THE STRAIN

'Deben' - the latest addition to the Alexandra fleet. Launched by H.R.H. Princess Alexandra on 25th November, 1989. Powered by 3860 h.p. and with stern drive azimuthing propellers, she has a bollard pull of 46 tonnes. She is now in service in Felixstowe - one of Europe's busiest ports.

TAKE THE STRAIN

The Alexandra Towing Company and the British Tugboat Business, 1833-1987

TIM NICHOLSON

The Alexandra Towing Co. Ltd

End Papers — Ackerman print of the Port of Liverpool, mid 19th century in the possession of Fritz Spiegl. Liverpool Packet No. 2 + 2.

First published 1990 by
The Alexandra Towing Co. Ltd
Castle Chambers
43 Castle Street
Liverpool L2 9TA

Publishing Consultant: K. Allan Brunton-Reed,
Thomas Reed Publications Ltd

ISBN 0 9516495 0 7

Typeset in Palatino 10/12½ pt by MHL Typesetting Ltd, Coventry

Printed in Great Britain by
Redwood Press Ltd, Melksham

Contents

Acknowledgements

The author offers grateful thanks to the following, for help and advice freely given:

R.L. Adam
Robin Craig
Richard A. Foster
A.J. Francis, Naval Historical Library, Ministry of Defence
Richard Martin, Mersey Docks & Harbour Company
Nigel Overton, Tudor House Museum, Southampton
Miss D. Procter
J. Gordon Read
Janet Smith, Liverpool Record Office
M.K. Stammers, Merseyside Maritime Museum
Eric Taplin and Tony Lane, Department of Economic History, University of Liverpool
Fred M. Walker, National Maritime Museum
Dick Williams, Associated British Ports, Swansea and Port Talbot

It is not invidious to single out three people in particular, for their unstinting patience with what must at times have seemed an endless stream of questions:

Dr P.N. Davies, Department of Economic History, University of Liverpool
Phil Thomas
Miss Elizabeth Wiggans, National Maritime Museum

Public and corporate bodies are among those without whose help this book would have been impossible to write:

Associated British Ports, Garston
Central Statistical Office
City of Glasgow Libraries Department
Hampshire Divisional Library, Southampton
Leisure Services Department, Humberside County Council
Local Studies Department, Newcastle upon Tyne Central Library
Maritime Museum, Borough of Hartlepool
Middlesbrough Central Reference Library
Museum of Transport, Glasgow
Ocean Transport and Trading Plc
Rea Towing Co Ltd
Rochester County Library
RTZ Ltd
Shell UK Ltd
Town Docks Museum, Kingston upon Hull
Transport General Workers' Union, Research Department
Trustees of the National Museums & Galleries on Merseyside

Serving directors and staff of The Alexandra Towing Company and its subsidiaries have been uniformly helpful, courteous, and generous with their time and their precious records. Among them were -

Henry Bicket, Graham Mack, Alastair Mack, Mike Banbury, Dick Hazell, Colin Butt, Ces Smith, Brian Clarke, Graham Gibbons, Gerry Gilbert, Graham Gollar, and the much put-upon but unfailingly patient Ivy Bryce and Rose Bartley.

The author remembers with especial gratitude Alexandra's 'floating' staff for their hospitality and forbearance on the bridges of their boats, particularly:

Brian Biggs, Bob Bladon, Arthur Langton, Bob Radford, John Reynolds, and Jack Simmons.

With their reminiscences, the retired directors and staff of Alexandra and its associated companies kindly endowed the book with any colour it may possess:

Col. A.J. Page, G.R.D. Scott, Robert Air, Ron Bullen, T.H. Richards, Arthur Maddocks, George Hilton, A. Dickson, Gareth Mills, Bill Beatty.

The publishers would like to express their gratitude to Alastair Mack, Brian Clarke and Ken Troup for their contribution to this book. They were responsible for the difficult task of selecting all the photographs, and also saw the book through its various stages of production.

The Alexandra Group of Companies

The Alexandra Towing Company Limited
The Alexandra Towing Co. (London) Limited
    Britannia Marine Towing Equipment Limited
    The Medway Dry Dock and Engineering Co. Limited
The Alexandra Towing Co. (Felixstowe) Limited
Bulk Cargo Handling Services Limited
    Trafford Park Stevedoring Company Limited
    Alexandra Molyneux Haulage Limited
Alexandra Aviation Limited
    Euroair Transport Limited
Alexandra Marine Transportation Limited
The Royal Terrace Pier Estate Company Limited
B.A.C. Group Limited (Associate Company)

Past Chairmen

William Poulsom Esq., J.P.
1887–1903

John Griffith Hughes Esq., J.P.
1903–1909

James Jermyn Mack Esq. J.P.
1909–1933

Alexander Bicket Esq.
1933–1967

Harold Brooke Mack Esq.,
1967–1972

Henry Brussel Bicket Esq. J.P., D.L.
1972–1988

Foreword

Harvey Graham Mack Esq.,

The Alexandra Towing Company Limited was incorporated as a Company Limited by shares on 11 August 1887 under the Companies Acts of 1862 to 1886, so 1987 was its Centenary Year. The present Board of Directors decided to mark the occasion by setting up an Exhibition to show, in a small way, how the Company's tugs had developed over the years and the part they had played in the development of the towage industry throughout the momentous years from the end of the nineteenth century to near the end of the twentieth.

Exhibitions, however good, are necessarily transitory, so the Board decided to commission a book to place it all on record permanently. They decided that this book should embrace the changing pattern of political and commercial life, particularly ports and shipping and the way in which the towage industry, in particular The Alexandra Towing Company, adapted and changed. This was a difficult task because during the century, over two hundred tugs varying greatly in size and power passed through its hands and two World wars had occurred with all their devastating effects. However Tim Nicholson has brought it all together into a wide ranging record of a hundred and fifty years of the British tugboat business which includes the first century of the life and times of The Alexandra Towing Co Ltd.

The book relates how, over the years, the Company has expanded by acquisition and from its original port of Liverpool it now operates in London,

Felixstowe, Southampton, Swansea, Port Talbot, Barrow and Gibraltar. It owns and operates a fleet of modern, high powered, harbour and estuarial tugs second to none, with two in course of construction and more on the drawing board. At the same time, it is involved in offshore and deep sea towage, and the Company's business interests include aircraft ownership, leasing, brokerage and road haulage. A considerable change from six small tugs in 1887.

The developments could not, of course, have been achieved without the dedication, skill, and determination of our predecessors and those who follow them in all sections of the Company, especially crews and management.

The future cannot be foreseen but we are proud of the achievements of the first hundred years and hopefully this book will be read by many for both its historical and human interest.

I hope all who read the book will enjoy it.

H. Graham Mack
Chairman

Bow Lane
London
June 1990

Deben

Charlotte Dundas

Introduction

The tug moves ships. Left on their own in docks and estuaries, many of the biggest ships that feed Britain, take out its exports and bring in its imports must flounder helpless, a menace to themselves and everything around them.

The tug moves lighters. Pushed and pulled by tugs, these floating boxes carry cargo from ship to shore where docks and quays are shallow, small, or non-existent. When they bring coal and raw materials to power stations, chemical plants, and paper mills on rivers too shallow or narrow for ships, they not only bring power, heat, and light to millions of people and thousands of businesses; they also bring wealth and jobs.

The tug is in at the birth and the death of a ship. It takes charge as soon as the new vessel hits the water, still an inert hulk with no life of its own, and tows it to its fitting-out yard. When the ship has reached the end of its useful life, the tug tows it to the breakers. Nothing else can do the job.

The tug moves oil production platforms, the size of cathedrals and about as seaworthy, on pontoons to their offshore oilfields. Before the days of the heavy lift vessel - a sort of powered raft - there was no other way. No tugs, no North Sea oil, when Britain most needed it.

The tug rescues the multi-million-pound ship, stranded on a sandbank or run ashore - perhaps because its master thought he could do without a tug.

The tug saves lives - ask the six men from the ferry *European Gateway*, fished from the black North Sea on a freezing December night in 1982 by *Ganges, Alfred,* and *Sauria* out of Felixstowe. The tug, often first on the scene with the equipment needed, fights fires on ships and shore, and fights oil spills from tankers. *Torrey Canyon* and *Pacific Glory* are among its battle honours.

The tug tows out a city's refuse for dumping at sea. With a little smartening-up, the same tug has served as a yacht for the rich, cheerful, and drunk, and as a pleasure boat for the poor, cheerful, and drunk. It has seen home Cleopatra's Needle, Channel swimmers, chunks of D-Day's

Mulberry Harbour, half a tanker, and rather less than half of Henry VIII's warship *Mary Rose*. Unofficially, one tug has been a floating maternity ward; and another has gone whooping off on a whale hunt.

The tug has been, and is, many things to many men. Yet at best, it is patronized. The layman's eye sees selectively, and in journalists' clichés. A passenger liner, usually an 'ocean giant' and usually departing, occupies centre stage as it sets out on the Psalmist's lordly, if vague, 'business on great waters'. The tugs are always the bridesmaids, never the bride. Their exact function left just as vague, they are amusing little craft 'fussing around' the receding monster. Joseph Conrad used such words to describe the exit of a clipper from a London dock: 'A tug or two would hurry in noisily, hovering round her with an air of fuss and solicitude, and take her out into the river, shepherding her'. It was the clipper or the big steamer, not the tug, that interested him.

At worst, the tug has been despised. It was so in J.M.W. Turner's 1838 painting of *The Fighting Temeraire*; a nasty, grubby symbol of modern 'smoke, soot, iron, and steam'. Turner contrasted it deliberately with the romantic image of a symbol of Britain's past glory, fading in a haze of rampant nostalgia against the setting sun.

The tug may no longer attract such active hostility; but if it is not patronized, it is ignored. The reason is in part simply physical: the tug is small, and when it tows a ship, its charge, as a rule, is overpoweringly bigger. Partly it is that, instead of enjoying an obviously independent character, the tug exists only to provide a service to another business, and one of overshadowing proportions - the commerce of a nation.

The arena of this service is the port. It makes no sense to talk about tugs except in the context of their ports. Their working day usually begins and ends in ports. They would not exist but for the business generated by ports. They are part of the machinery of a port, reacting to the demands of ship movements within it. Because the fortunes of ports and tugs are so intimately linked, the ups and downs of the one are faithfully reflected in those of the other. By enabling a port to handle bigger ships than was possible before the tug existed, and hence to handle more freight, it contributed in the last century to that port's growth and prosperity. In doing so it performed the same service for the nation, for ports are the channels through which commerce keeps a country alive.

Today's Alexandra Towing Company has absorbed, over the past hundred years, businesses that between them have operated from almost the earliest days of tugs until the present. Between them they have worked in five of Britain's major ports. Two - London and Liverpool - were and are the first and second in terms of business done. All the rest - Swansea, Southampton, Felixstowe - have ranked high in the league table in the past, or do so now. No picture of these ports alone can tell the complete story of Britain's outlets

'Brockenhurst' has a line aboard the 119,000 D.W.T. 'Pacific Glory'; after she collided with 'Allegro' off the Isle of Wight, and then ran aground bursting into flames. (Chris Wood, Southampton).

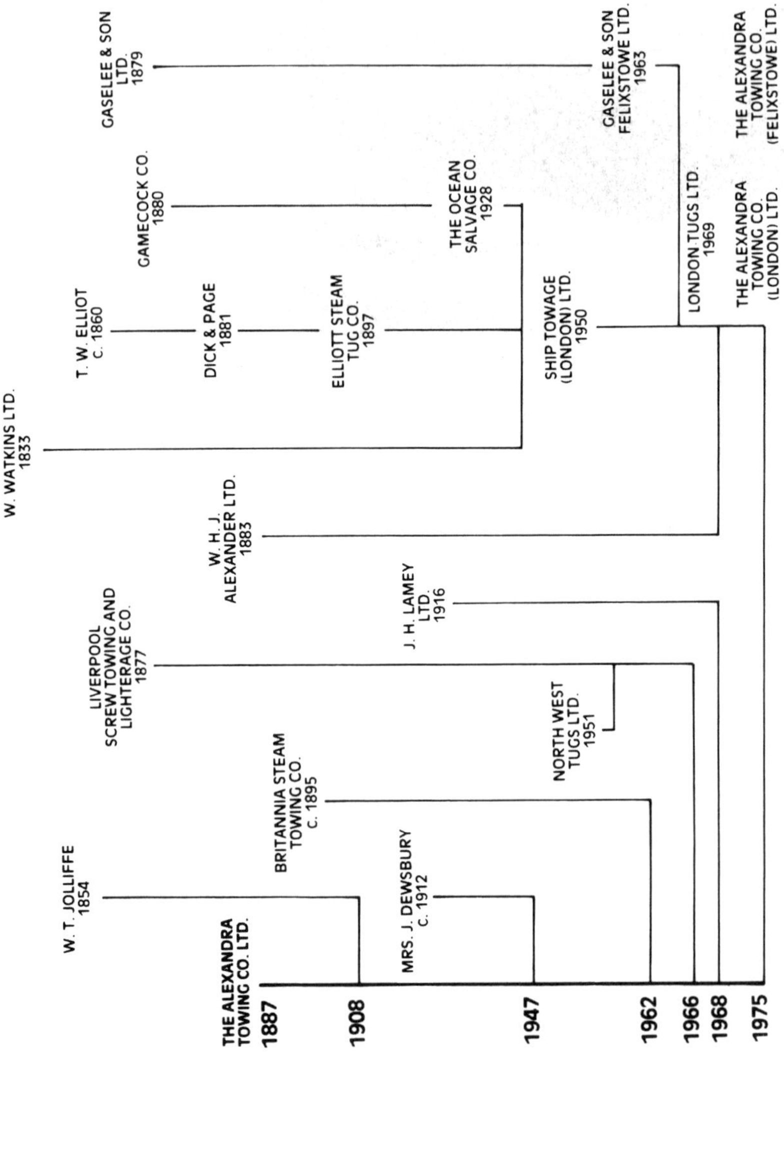
THE ALEXANDRA TOWING CO. LTD. 1887
1908
1947
1962
1966
1968
1975
W. T. JOLLIFFE 1854
MRS. J. DEWSBURY C. 1912
BRITANNIA STEAM TOWING CO. C. 1895
LIVERPOOL SCREW TOWING AND LIGHTERAGE CO. 1877
NORTH WEST TUGS LTD. 1951
J. H. LAMEY LTD. 1916
W. H. J. ALEXANDER LTD. 1883
W. WATKINS LTD. 1833
T. W. ELLIOT C. 1860
DICK & PAGE 1881
ELLIOTT STEAM TUG CO. 1897
SHIP TOWAGE (LONDON) LTD. 1950
GAMECOCK CO. 1880
THE OCEAN SALVAGE CO. 1928
LONDON TUGS LTD. 1969
THE ALEXANDRA TOWING CO. (LONDON) LTD.
GASELEE & SON LTD. 1879
GASELEE & SON FELIXSTOWE LTD. 1963
THE ALEXANDRA TOWING CO. (FELIXSTOWE) LTD.

Family Tree

to the world, but it does give a representative one.

The story that follows must, it is true, begin with waterways, ports and tugs that the company's constituent operators never knew. The very earliest tugs ventured out on to Britain's waterways a generation before the first 'Alexandra' company - as we shall call them generically - joined in. But thereafter the story will stay with Alexandra company ports, visiting them as new enterprises come to life, returning to them to catch up with events when other Alexandra businesses start up. So, too, with the tug operators who make up the Alexandra family tree on page 4. There have been hundreds of other firms, perhaps thousands of tugs; but the scope and scale of the work of the dozen tug owners and their four hundred or so boats, most of them prominent and busy operators in their time, over a period of more than a century and a half make their story a very fair microcosm of the whole. This is doubly so when the different companies' competitors in a port come into the picture as well, as they must, for without them, the behaviour of our main protagonists does not make sense.

A story must have a framework, and large, rather remote elements - ports and companies - provide the only possible one here. But the machines and the men who work them give the tale its movement and colour. The mechanical principles and technical advances that moulded the tug must be understood, if only in outline, so that its capabilities and limitations may also be understood.

These technical developments were common to many ports and operators. So, too, were the circumstances of the tugman's life. There was the grand, romantic adventure of 'seeking' the homebound sailing ships, in which steam and sail became for a couple of generations interdependent, before the one destroyed the other. In contrast, there was, and is, the hard reality of the tugman's everyday life; in one family of a father and four sons, two were killed and two more crippled in separate, unrelated accidents between the 1920s and the 1950s.

Death can come by the decapitating whip of a parted towing line; by obliteration when a tow perhaps a hundred times a tug's weight overruns her, ploughing her into the sea; by 'girting' as a dead ship's drift accelerates, pulling the tug over sideways and tipping her crew into the water. But the tugman cannot let thoughts of such things rule his life. Instead, he lived, and to a great extent still lives, with the unrelenting strain of intense competition with other tugs for the same jobs, of being on call around the clock to meet demand whenever a job arose, and then of working a twenty-four hour day or a seven-day week whenever the work demanded it. When trade was slack, jobs consequently scarce, and rivalry for them correspondingly fierce, operators were sometimes too poor to repair, let alone replace, a worn-out tug. Men might be sent to sea in leaking, rusting deathtraps, with which they were expected to win a race for a tow or a salvage job, then fight their way back to port with their frequently helpless, always unwieldy charge, through any kind of weather. From time to time a tug simply disappeared, leaving not a ripple.

Small wonder that one tugmaster bombarded a competitor with abuse and empty bottles, another rammed a rival, and a third rammed the customer who had rejected his services. Yet another, on being presented with a pirate's Jolly Roger by his resentful but - one feels - half-admiring competitors, ran it up proudly at his masthead. Provisioned with full bottles this time, a third tug's crew, who could not have numbered more than nine men, turned Ramsey harbour into a shambles of broken bones, bloody heads, and listing, drifting small craft.

There was, finally, the shared horror of the two world wars, in which tugs usually too insignificant to be worth a torpedo, were nevertheless machine-gunned to splinters in the Thames estuary, blown up by mines in the Mersey, dive-bombed at Dunkirk, run down on dark Channel nights by friendly warships, and even, on one occasion, demolished by a VI flying bomb.

Men, tugs, companies, and ports come together as the ingredients of a story that combines enterprise with complacency, inventiveness with blind conservatism, high drama with low comedy, and courage with the everyday hard work of men in a business that has always been a commercial gamble, demanding every sort of physical and mental endurance.

1

Monarch

Beginnings

When the Cornish engineer Thomas Newcomen made the earliest practical reciprocating steam engine in 1712, he gave the world its first artificial power source, independent of natural forces such as wind, water, and human or animal muscle. In its original form the steam engine was a massive structure, more like a building than a machine, set into the ground on stone foundations and used mainly to pump underground water from coal mines. Because it made steam at a pressure of (at most) a few pounds per square inch above that of the atmosphere, the boiler could be made of wood, like much of the rest of the engine. Its steam was admitted to the base of a vertical iron cylinder at least six to eight feet tall, two feet in diameter, and perhaps several tons in weight.

In theory, the engine was extremely inefficient. The steam in the cylinder was condensed by an injection of cold water - which cooled the cylinder too, wasting heat and energy. The resulting vacuum under the piston allowed atmospheric pressure to force the piston down. The piston was suspended at one end of an overhead rocking beam, on the other end of which hung the pump rod. The counterweight of the rod caused the piston to rise. The cycle thus completed was repeated not more than a dozen times a minute; a rate that was, in practice, about double the usual working speed of the engine.

Primitive though it was, a low-pressure condensing engine transformed the productivity of the coal industry by allowing deeper mines than ever before to be exploited to feed iron foundries, glassworks, brickworks, soap boiling, salt extraction, and other manufacturing industries dependent on fuel, at a time when charcoal from Britain's disappearing forests was becoming scarce and dear. One pillar supporting Britain's coming Industrial Revolution had been set up.

If an artificial power source could be substituted for natural ones to such effect in one environment, why not in another, where its potential was at least as apparent? In the eighteenth century sailing ships were the world's prime international cargo carriers; yet they were wholly dependent on wind, tide, and muscle. Wind was unpredictable; and without it sail was helpless against contrary tides and currents. When the wind died a sailing ship stopped, perhaps within sight of harbour, and might not move again for weeks. Even if the wind stayed favourable, a ship would have to take in sail in the approaches to a port, which were likely to be crowded with other craft, and full of hidden shoals and winding, narrow channels. Before this point was reached, the land closing in about a ship might take the wind out of her sails. In any event, she stopped well short of her destination at the port's wharf.

In small ships sweeps - long oars - could then be unshipped. A bigger vessel could keep going for a while by putting her men over the side in the ship's rowing boats to tow her. But the use of oars was limited by the endurance of the men pulling them; and oarsmen could not make way against a strong wind or tide. In any case, there was a limit on the size of ships that could be pulled by rowing boats, even in a calm, and even if other ship's crews and local watermen were hired to help. Once a ship was close to land, a line could be taken ashore, there to be picked up by men or horses, or fastened to a bollard so that the ship's crew on their capstan could warp the vessel into its berth. When the approach to a river port was long and devious, or when the channel and the water at the port's wharfs were not deep enough for a big ship, the vessel moored in the nearest deep water, and its cargo was offloaded into shallow-draught lighters and rowed, sailed, or pulled by horses upstream to the port.

It was natural enough that when inventors came to consider how steam could be applied to purposes other than pumping, a self-propelled ship for towing other craft should spring to mind early on. Twenty-four years after Newcomen built his first pumping engine, an English clock repairer, Jonathan Hulls, made the necessary imaginative leap. His patent of 1736 proposed a boat driven by a Newcomen low-pressure condensing engine that turned a single stern paddle-wheel. Hulls's 'New Machine' was designed for 'carrying vessels or ships out of or into any harbour, port or river, against wind or tide, or in a calm'.

To be usable for driving any sort of rotating machinery, whether of a factory machine or of a self-propelled boat, a power source had to be directed to turning a shaft with wheels on it. That is to say, the reciprocating, or to-and-fro, motion of the steam engine's piston had to be converted into rotary motion. The French physicist Denis Papin had devised a ratchet and pawl mechanism for doing this, probably for a model of a steam road vehicle he made in 1698. It reappeared in Hulls's patent as part of a system by which the engine turned the paddle-wheel.

The paddle-wheel, an ancient idea deriving from water-mills, had featured in a Papin proposal of 1690 for a boat driven by a reciprocating steam-engine. In its most elementary form, it consisted of a wheel with blades called floats fixed around its circumference. The floats entered the water at an angle to its flow, pushing it backwards and the boat forwards.

Neither Hulls's nor any other steamboat progressed beyond an idea for nearly half a century, because of the unsuitability of the existing, ponderous Newcomen engine for any but terrestrial use. But by the mid-1780s the engineers John Smeaton and James Watt had so refined the original steam engine that it had become eight times as efficient. Watt's separate condenser and steam jacket for the cylinder cut down heat loss, thereby conserving energy and reducing fuel consumption. Watt had devised the double-acting engine, in which steam at low pressure admitted at the top of the cylinder helped atmospheric pressure to drive the piston down, and then incoming low-pressure steam at the bottom of the cylinder pushed it up. For the first time, steam was being used directly to propel the piston. This innovation alone doubled an engine's power output. Furthermore Watt's engines had flywheels, which allowed power to flow out smoothly and consistently.

Two more methods of turning reciprocating into rotary motion were now available, both of them more efficient than the ratchet and pawl. Watt enabled his engines to turn shafts and wheels by means of sun and planet gearing, while in 1779, James Pickard, a Birmingham button manufacturer, patented a means of imparting rotary motion with a crankshaft.

These improvements were initially applied only to manufacturing industry, where they turned steam into an efficient and economic proposition for such innovators as Matthew Boulton, a Birmingham manufacturer of miscellaneous metal goods. But many were seized on by steamboat experimenters as they became known, for their usefulness was even greater on water than on land. A boat engine had to be reasonably small and light, yet produce enough power to propel the weight of the vessel against the resistance of the water. The more efficient a steam engine, the more compact it could be for the same power output. Systems of conveying the engine's power to a turning shaft were as important to a steamboat designer as they were to Pickard and Boulton. The low-pressure engine was still feeble for its size and weight; but it now had the ability to propel a boat successfully - if only, as yet, on sheltered waters such as rivers, lakes, and canals.

The canal boat did not have the problems of ocean-going sailing ships - or even of river boats - with strong tides or currents. It navigated the narrowest and most sheltered waters of all. It was sailed or drawn from the bank by horses; and it had no docking problems. But a sailing boat on a narrow waterway was particularly vunerable to contrary winds; and muscle had its limitations. Add to this the fact that the first steamboats were frail craft normally developing only 4 or 5 hp, and it is not surprising that the first vessels to be towed by steam were canal boats - burdens not much greater than the towing vessel itself.

In 1789 the Forth and Clyde Canal was almost completed. It rescued traders and manufacturers from the bad roads that crossed the waist of Scotland, and from the only alternative - the long sea passage around Cape Wrath. When finished two years later it would make a continuous 39-mile waterway between Grangemouth on the Firth of Forth and the Clyde, with a branch south to Port Dundas joining it to the Monkland Canal, which served the growing industries of west central Scotland. It could take sailing boats of up to 100 tons and 7½ ft draught; but unfavourable winds meant delays and congestion. The first steam experiment on the Forth and Clyde Canal took place in December 1789, when the Scots engineer William Symington, backed financially by the landowner and amateur engineer Patrick Miller, launched a boat with a two-cylinder engine. It was intended for 'dragging vessels', but did not get further than an unaccompanied demonstration, when it achieved a speed of about 6 knots. Miller withdrew his financial support from Symington, who made no more steamboats for a dozen years.

The ten-mile Bridgewater Canal had been opened in 1761 to allow coal from the Duke of Bridgewater's mines at Worsley to be carried easily and

'Charlotte Dundas' designed by William Symington and named after the daughter of his sponsor Lord Dundas. (National Maritime Museum — Greenwich).

cheaply to its Manchester market. Between 1796 and 1799 the Duke tried out steamboats on the canal, possibly towing coal barges; but the wash from their paddles damaged the canal bottom and banks.

The earliest trials of a tugboat to be accounted a success took place further north two years later. By 1801 the Forth and Clyde Canal had been open throughout its length for a decade. In that year the first steamboat to be christened *Charlotte Dundas*, designed by William Symington, was completed at Grangemouth with the help of a new backer - Thomas, Lord Dundas, a governor of the canal company. She was named after his daughter. Trials were held on the Forth and Carron rivers, but even though she was 'running free' - not towing - she lacked power, and was abandoned.

Symington's second *Charlotte Dundas*, launched in 1802, remedied the deficiency: her double-acting engine developed a remarkable 10 hp. She was revolutionary in another respect, too. Conventional low-pressure engines, with their rocking beam above a vertical cylinder, had too high a centre of gravity for stability in a boat, and took up a lot of headroom. Symington's solution was to mount the cylinder horizontally. Drive was through a crankshaft to a stern paddle-wheel, which was geared up to reduce its speed, thus - it was hoped - reducing water turbulence and damage to the canal. Engine and boiler were exposed to the open air, as was normal in the earliest steamboats. There was a double stern and rudders on each side of the central paddle-wheel. The steersman stood in the bows, for the paddle-box would have blocked his view from the stern. The vessel was 56 ft long and 18 ft in the beam.

In March 1802 the second *Charlotte Dundas* drew two laden vessels weighing 70 tons each from Lock 20 on the Forth and Clyde Canal to Port Dundas, making the 18½-mile journey in 9¼ hours. But damage to the canal banks from wash meant that the experiment was not repeated; and the boat was broken up. Steam did not return to the Forth and Clyde Canal for another twenty years. The third Duke of Bridgewater ordered a fleet of eight Symington steamers for his canal, but he died, and they were never delivered.

Although canal interests had not taken to the steamboat, in spite of its apparent advantages, the impetus behind its development - and hence of the tug - remained in the north of Britain. At the western end of the new inter-ocean waterway, Glasgow was being developed as a port. Although the Clyde was broad, until the early nineteenth century, shoals prevented sea-going ships from sailing higher than Port Glasgow, twenty miles downstream. There the anchorage was congested with vessels discharging their cargoes either into lighters towed by horses for the onward journey upstream, or else on to packhorses for overland carriage. But Glasgow's coal and iron industries were booming - 1782 saw the opening of the Clyde Iron Works - and the town wanted ships to discharge on its doorstep.

Deepening the river by dredging began, enabling a 120-ton ship drawing 8½ feet of water to reach Glasgow by 1806, and accelerating after that date.

An increase in Clyde-owned shipping attracted shipbuilders to the river - most momentously, builders of steamships. At Port Glasgow a Scots hotelier, Henry Bell, built his *Comet*, which made its first trip in August 1812. She was designed as a passenger steamer, plying between Glasgow and Greenock. Here and on other rivers lined with increasingly populous towns, the passenger trade at first looked likely to offer the best returns.

The 28-ton *Comet* incorporated a major change in steamer design that became universal in paddle tugs, too. Most steam engine designers were conservative folk. Bell's solution to the problem of a high centre of gravity in steam engines was to retain the established vertical layout while getting rid of the overhead rocking beam. Instead, piston rods worked two short levers set low down, one each side of the 4 hp engine, which in turn connected rods linked to a crankshaft. This turned a pair of side paddle-wheels; which was not a new idea. In a tug the arrangement would be more practical than a single stern wheel, which interfered with towing. Unlike the second *Charlotte Dundas*, the *Comet* was rigged for sail assistance. This was another characteristic of early steamboats, including most tugs. Sail conserved fuel, made up for lack of power, and was a lifeline in case of breakdown.

Four more steamers were built on the Clyde in 1813, nine in 1814, eight in 1816, and forty eight in 1822. Glasgow became the main centre of steamship building in Britain, a development which transformed it into a great industrial city. It comes as no surprise to find, as early as 1814, that steamboats that had been carrying cargo between Greenock and Glasgow were now also towing lighters on the same route. There was continuing need for their services: the deepening and otherwise improving of the Clyde for navigation continued for another century. An instructive case history was that of the *Industry*, built as a passenger steamer in 1814, then converted to carry cargo as well, and finally set to towing lighters.

Most early steamboats could take on a variety of roles without much modification. All manner of steamers acted as tugs when required, so long as the towed vessel - the 'tow' - was of a manageable size. But in 1817 one was built at Dumbarton on the Clyde specifically for towing on the Firth of Forth between Grangemouth and Leith, the port for Edinburgh. She was named *Tug*: the earliest known use of the word for a specialized type of steamboat. A year later a new vessel joined the sixteen other steamers by now on the Clyde. Built for the Clyde Shipping Company, which already owned the *Industry*, she was aptly named *Samson*. Although intended as a passenger steamer as well as a tug, her two 20 hp engines showed her designer's foresight.

In the north-east of England, Newcastle upon Tyne was already a flourishing port. It exported coal, mostly coastwise to the insatiable and multiplying fireplaces and furnaces of London - 680,000 tons a year by 1775. Newcastle was a one-industry town, and the mouth of the Tyne had a difficult exit: when the wind was easterly the London collier fleet could be held up for weeks, a few ships (the first out) making a fortune while the rest lost time and money. Furthermore, no incoming merchant ship bigger than 240 tons tried to come as far up river as Newcastle; their cargoes were expensively transhipped at Shields.

Like the Clyde, the Tyne seemed tailor-made for steam towing; but as on the Clyde, the first steamer on the river was built for the passenger trade. She was the *Tyne Steamboat*, launched at Gateshead in February 1814, and plying between Newcastle and Shields. But she did not pay as a passenger steamer, and after four years, renamed *Perseverance*, she was turned over to towing. Her first task in her new role was to tow *The Friends' Adventure*

Two wooden paddle tugs similar to William Watkins' first tug 'Monarch', 1833, made famous in J.M.W. Turner's painting 'The Fighting Temeraire', photographed by Rev Calvert Jones in Swansea Harbour in the 1840s.

from Newcastle two miles out to sea. The introduction of steam towing increased the number of sailings of the London collier fleet from eight to thirteen or fourteen a year; and with its help merchant ships of up to 400 tons now penetrated upstream to Newcastle. By 1821 there were fourteen steamboats working on the Tyne.

No one knows how many steamers were engaged in towing ships, which was to become the tug's main function in life; nor how often they did so. At this time the future of steam in ship towing was still uncertain. It was not merely that steamboats were usually low-powered, unable to make headway against strong tides or currents, and able to tow only in calm weather. A steam tug was not yet thought indispensable for ship handling. Most ships were not yet too big to be towed by muscle power, which was free if the ship's own crew were used, and at worst was cheap - certainly cheaper than 'taking steam'. Nor, in the days of sail, was speed normally regarded as a necessity. Voyages were long in any case, so there was usually no special rush to shift a ship in or out of port. Muscle power did the job eventually, if slowly.

It seemed that there was no guaranteed future even for lighter towing, in spite of the apparent market for it. On the Clyde and the Tyne most entrepreneurs thought of steamers as passenger boats first, cargo boats second, and tugs last. As well as switching roles the other way, steamers used as tugs, having lost money, were turned over to passenger work. The most salutory, not to say ironic, fate awaited the original *Tug*. Sold off, she was converted as late as 1838 - into a sailing ship. Nor was she the only one.

In the 1820s the Admiralty alone had made up its mind. By 1822 it had bought one steamer (the *Monkey*), and ordered another to serve as tugs for towing ships of the line out of harbour when winds were unfavourable. If a national emergency arose, speed would be more important than cost-effectiveness. In the world of commerce the idea of the steamboat as a tug had to wait until the following decade before it became accepted, first on the River Thames, then on the Mersey.

2

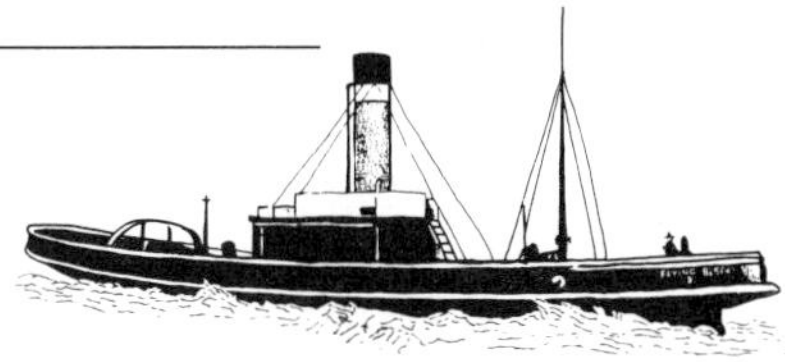

Flying Breeze 1882

Early Developments: London and Liverpool

In the 1820s London had been Britain's busiest and richest port for upwards of a hundred years, in terms of the tonnage of shipping and the quantity and value of goods it handled, in spite of the fact that unlike Liverpool or Swansea, Glasgow or Tyneside, it had no adjacent industrial or manufacturing region to provide its exports or absorb its imports. London was primarily an entrepot - a funnel through which imported goods were disseminated throughout the country, or trans-shipped abroad as re-exports. Its communications were second to none. Rivers, canals, and roads radiated from it; within twenty years it would be the hub of a nationwide railway system as well. Conveniently placed for continental and many other British ports, London offered ships easy access from the sea and ample anchorage. It was situated on a long, sheltered, tidal river deep enough for the merchant ship of the day, with a mouth unencumbered by a bar, the submerged sandbank that obstructed so many estuaries. Unlike other growing ports, its river would need no great improvements, other than occasional dredging of channels, for almost a century. And although London served no immediate industrial area, it had been for centuries the heart of government, finance, commerce, and culture, with all the prestige of a capital city. It had become a magnet for population, which grew from 1.1 million to 2.7 million in the single generation 1820-51. It therefore absorbed huge quantities of food, fuel, construction materials, and other necessities, as well as distributing them.

Until the beginning of the century there had been only one 'wet' dock on the Thames; that at Rotherhithe, opened in 1700. A wet dock was one permanently filled with water (unlike a dry dock) where ships moored while they discharged or took on cargo. On a tidal river such as the Thames, ships entered wet docks through locks, which kept the water within at a constant

The first 'wet' dock in London was at Rotherhithe, opened in 1700. (National Maritime Museum — Greenwich).

level, insulated from the rise and fall of the tide outside. On the Thames this could be 16 to 18 ft. Where wet docks were lacking, ships had to discharge their cargoes at riverside wharfs, hampered by changing water levels, or into lighters while at anchor in the river. On the Thames ships could be unloaded only at 'legal' wharfs, whose owners had a licensed monopoly of the facilities, and therefore an interest in restricting their extent. Queues of ships formed. Discharging cargo into lighters in mid-river caused further congestion.

Twenty-two years of war with France brought a new demand for accommodation and shelter for both merchantmen and warships, and the bottleneck burst. On the north bank the West India Docks were opened in 1802, and the construction of the East India Docks began in the following year. The 38 acres of the London Docks were added in 1805. On the Surrey shore the Commercial Dock Company opened new docks at Rotherhithe in 1807.

New docks increased business for lightermen rather than reducing it. Because the docks were spread out along an ever-lengthening stretch of river, goods were more easily and cheaply transferred from dock to customer by river on lighters than through crowded streets. Furthermore, many ships continued to discharge in mid-stream, to avoid dock charges. Even so, the Thames lightermen were compensated for their anticipated loss of business by being allowed to load and unload ships in the docks free of charge. The lighterage business flourished.

As some of their names suggested, the dock companies enjoyed monopolies of particular trades: ships engaged in them could use no other docks. Wartime expansion ended with Napoleon's defeat; but starting in 1823, dock companies' monopolies were terminated when they came up for renewal. This had the effect of encouraging other builders to construct docks. In 1828, St Katharine's Dock was opened, adding 10½ acres to the facilities; and in 1835, the Victoria Dock. With the docks spread the new bonded warehouses, where dutiable goods could be stored in bond until they were re-exported or otherwise sold away from the river.

At about this time Britain's merchant marine numbered around 19,000 ships, totalling between 2.2 and 2.4 million tons. No less than a quarter of this tonnage (555,000 - 600,000 tons) was registered in London. Newcastle upon Tyne came a long way second, with 202,000 tons - mainly colliers - and all other ports had still less. Ships owned in London in 1829 averaged 215 tons. This figure reflects the small size of most ships - ocean-going merchantmen were more likely to be 700 - 800 tonners - but it was still nearly double the national average: a reflection, perhaps, of the ease of the Thames approach, and of the city's dock development.

The growth of shipping on the Thames reflected the growth of trade. The coastwise coal trade from the north-eastern ports was huge: two million tons a year was coming into the Thames by 1830. In the country as a whole, but in London in particular, shipping was being encouraged by the end of the East India Company's monopolies. In 1814 its monopoly of trade with India and the East Indies had been abolished; from 1834 its last remaining monopoly, of trade with China, was no more. Firms trading with both regions multiplied. By 1816 the volume of British shipping trading to the East Indies was four times that of 1792. Most of this trade remained with London, where it had always been.

Although protection of British products by means of duties on foreign goods was still official policy, free trade in raw materials that Britain could not produce was increasingly a fact of life, under pressure of demand from her manufacturing industries. Not only London benefited. Raw cotton was virtually free from duty from the 1820s; the 100,000 tons a year imported into Liverpool had nearly doubled by the end of the following decade, and was 346,000 tons by 1847. Timber was needed for shipbuilding, regardless of duty. Two-thirds came from the colonies, but the rest was imported from Scandinavia and the Baltic. Timber was Britain's biggest import in volume and value, and by 1846 perhaps a third of the shipping tonnage entering Britain's ports was engaged in the timber trade. Exports boomed as well. Nationwide, the value of all exports rose from an average of £35.6 million a year in the late 1820s to £61 million twenty years later. All these goods, incoming and outgoing, called for ships to carry them; and the Thames had the lion's share of both trade and shipping.

These ships were not only British: more and more were foreign. For nearly 200 years the Navigation Acts had laid down that all goods imported into Britain and her colonies must be carried in British-owned ships, the crews of which must be at least three-quarters British. The object was to foster domestic shipbuilding and the maintenance of a large body of trained seamen for use in time of war; but the effect had been to discourage the carriage of British goods in foreign ships in many parts of the world, and generally to hamper trade. A new Navigation Act of 1825 began a process of relaxation. From now on European and, to a limited extent, other foreign ships could bring in the produce of their own countries. The ex-East India Company ships were among those now competing with one another in private hands. They were of prime quality - the wealthy Company had never had to cut corners - so their useful life would be long. They were big, too - up to 1400 tons - because they had to be able to defend themselves across the world's oceans, but the Company's monopolies had meant that they had no need for speed to outstrip competition. The waters on which the old East Indiamen were embarking were very different.

Now that there was competition, the aim was to cut down voyage time. For the first time they had to sail through the night instead of heaving to; and in a blow, they hung on to as much sail as possible for as long as they

Four twin-funnelled paddle steamers at the launch of 'Sans Pareil' from the Thames Iron Works, Blackwall, 9th May 1887. The only discernible tug name is 'Robert Bruce' on the right. (National Maritime Museum — Greenwich).

dared. Against them from the late 1830s, among others, were the new Thames-built Blackwall frigates in the Indian and Australian trades - big merchantmen, too, mostly of between 800 and 1400 tons, but designed on warship lines for speed, as their name suggested. Size was growing with speed, partly because the first was a condition of the second, and partly because the bigger the ship, the more economically it carried cargo. Size, however, brought its problems. Open-sea speed during a voyage was pointless if the 'warehouse to warehouse' time was long; and in a port or its approaches, it was harder and slower to navigate a big ship than a small one. There would be difficulties, too, for the watermen who pulled ships to and from their berths, and attended launches. The story was the same in trades other than those with India, China, and Australia. Commerce boomed from the 1830s; but so did the number, size and speed of the ships carrying it, and competition between them was intensifying.

The time was at last ripe for the steam tug on the Thames. In spite of the growing volume of trade and increasing size of ships, which would encourage the use of tugs, they came later here than on the Clyde or the Tyne, perhaps because London was more accessible to ships than were Glasgow or Newcastle. Bigger ships could make their way upriver more easily, reducing or postponing the need for tugs.

The first steamer to run regularly on the Thames was the *Marjory*, which appeared on the river in 1814. It is not known whether she was used for towing. But in August 1816 the steamboat *Majestic* is known to have towed the East Indiaman *Hope* from Deptford to Woolwich at an average speed of 2½ knots, against the wind. The earliest purpose-built tugs to work in the Thames did not arrive until 1832, when the Thames Towing Company brought the tiny, 18-ton *Lady Dundas* from the Tyne, and the 33-ton *Wear* from Sunderland. There was no local understanding or experience of what steam could, and could not, do: the *Lady Dundas* was so underpowered that she could not make headway against the Thames tide on her own, let alone with a ship in tow. Her owners got rid of her quickly.

In spite of this inauspicious start, the potential offered by the increase in size of shipping on the river, and the new call for haste, was too tempting to be ignored. Few men can have been better fitted to perceive it than John Rogers Watkins, 43 years of age in 1833, a Thames waterman of 23 years' standing who owned and operated both sailing ships and 'pulling boats'. That year his 14-year-old son William was apprenticed to the business, as his father had been 33 years before, and Watkins senior set up a company and named it after him. William's elder brother John, who had sea experience, ran the sailing-ship side of it. This was the first of the businesses that would eventually contribute to the make-up of The Alexandra Towing Company.

Convinced of the future of tugs on the Thames, John Rogers Watkins ordered one to be built for him on the Tyne. This was a serious step, for a waterman's business operated with little working capital, and a new vessel was a heavy investment. Whether a tug was new or old, running costs were high, engines unreliable and notorious coal guzzlers. There was no promise of success; the only guarantee was of competition, which was already there. Thomas Petley started to work tugs on the Thames at about the same time, buying second-hand and enjoying stronger financial backing. By 1836 at least two companies were established in rivalry with Watkins: Petley's Shipowners' Towing Company, and the Caledonian Steam Towing Company, with their new tug *Souter Johnny*. The pattern for the next century was set - one of small, usually under-capitalized, one-man or one-family businesses burdened with heavy running costs, highly vulnerable, competing for custom in a chancy market. Short-lived, seldom forseeable bursts of activity alternated with idle spells, during which tugs and crews ate overheads and contributed nothing to revenue while they waited for customers. To reduce overheads, most tug owners skippered their own tugs; and to get jobs they were always prepared to cut their prices.

But by building new, Watkins got a boat to his own specifications; and it turned out a great success. The 26-ton *Monarch*, nearly 65 ft long and 19 ft in the beam, had a 20 hp engine, which was power enough on the Thames in 1833. Her bunkers held 7 tons of coal: perhaps enough for 70 to 80 hours' running free, or less when she was racing for a job or straining with a tow. Coal consumption was a universal problem. On one occasion *Monarch* ran out of fuel while towing a collier to Gravesend, and had to 'borrow' some of its tow's cargo to keep going.

With her single engine turning both paddle-wheels together, *Monarch*, like other steamers of her day, would have lacked manoeuvrability had not Watkins pioneered a crude but effective remedy. An iron box filled with chain cable ran on rails athwartships. When a quick, tight turn was called for, the box was run to one side, its weight heeling the tug and lifting the paddle wheel on the other side out of the water so that it lost traction. The wheel still in the water bit deeper, spinning the tug on its axis. The system was adopted by all paddle tugs until the coming of separate, de-clutchable wheel shafts, and the more common double engines, each working one wheel.

Modern drawings and models show the *Monarch*'s towing hook situated just over half-way back from her bows, the situation from which it could most effectively pull a tow in a straight line, and turn it. This, too, was a feature eventually seen on nearly all tugs. *Monarch* was in service for 43 years, and it is possible that the hook was originally at the stern. She may have had an improvement that, by 1835, had been applied to the paddle-wheels of many steamers. This was the feathering float, which adjusted its angle automatically so as to enter and leave the water vertically. The old fixed float wasted power, and therefore fuel, by forcing the water down as it entered and lifting it up as it left. In the less widespread reefing float, the area of float presented to the water could, as the name suggests, be altered as the draught changed with the burning of coal, giving more or less traction as required.

Early tugs such as the *Monarch* presented no new problems to marine engineers or shipwrights. Conventional timber technology was adequate: tugs did not need to be heavily built, for great strength was not yet called for. The only major consideration for designers was a tug's vertical centre of gravity, which had to be low for lateral stability; but the paddle-boxes made it broad in the beam, and so naturally disinclined to roll. Towing was still a simple matter: it was a question of a single tug on a river or estuary, at the head of a string of lighters or pulling a ship of growing, but still manageable, size. That tugs were operating well within the limits of existing technology and techniques was illustrated by the lack of concern for their specifications shown by Lloyd's insurance underwriters, who laid down the type of timber to be used in their construction and the dimensions of their principal parts, but nothing else.

Monarch gained public attention not, of course, for her technical excellence, but because she was caught by J.M.W. Turner's pencil on 6 September, 1838, when engaged in a routine task: on charter to the Admiralty, towing a superannuated warship to the Rotherhithe ship-breakers. Turner and a friend, the sculptor W. F. Woodington, were returning up the Thames on a passenger steamer from Margate when he saw the ship appear 'in a great blazing sunset'. He promptly made a series of little sketches on the cards he carried for such occasions, later working them up into one of his most famous paintings, *The Fighting Temeraire*. His representation of the *Monarch* was inaccurate, placing her mast abaft her funnel, a mistake that was corrected by his engraver seven years later.

Intentionally, Turner's painting shows an ugly brute of a tug. In fact *Monarch* was better looking than that; and many early paddle tugs had truly elegant lines following those of crack sailing ships - slender hulls and clipper bows, flowing forward into long bowsprits. None the less, these were utilitarian and hard-working machines. It is almost possible, in the painting, to hear the churning paddle-wheels and smell the acrid coal smoke from the belching funnel. A spectator on board would have the reek of hot oil in his nostrils, and feel the deck trembling with the pounding thump of a loose-jointed side-lever engine. As a tug designed for sheltered waters turned into the wind, the deck under his feet would pitch in the chop, and salt spray sheet in over the low bulwarks. If a low-powered paddle-steamer was fighting an ebb tide as well, she might make no more than 2 or 3 knots even without a tow; a buoy moored in the channel would seem to take an age to creep past.

Watkins got an early share of work other than towing ships to the breakers. After the 600-ton steamer *Iberia* was launched in Limehouse in 1836, *Monarch* towed her to her fitting-out at East India Dock. This was one of the first steamships to be built for Arthur Anderson, soon to found the Peninsular and Oriental Steam Navigation Company. An important business connection had been established, for the P&O promised Watkins all their London towing work, a very early instance of a tug operator obtaining an exclusive contract from a shipowner. The P&O decided in 1840 to move to Southampton, but when they returned to the Thames in 1881, the link was faithfully renewed.

This was the kind of work at which the young William Watkins turned out to excel - the business side, which most tug operators, being ship handlers rather than businessmen by instinct, neglected to their cost. His father gave him much control early. In 1840, when he was 21, the company ordered a second tug, the more powerful *Fiddler*, this time built under its supervision on the Thames at Limehouse. In 1845 the *Lord Warden* was added and the *Monarch* was given a fire-tube boiler, rare at the time. In this design, the furnace gases passed in many small-bore tubes through the water, heating it. With its greater heating area, the fire-tube boiler raised steam more quickly than a conventional flue boiler in which the gases passed through a network of broad passages; it was more economical of fuel; and it allowed higher steam pressures (around 20 lb per square inch at this time), which in turn gave more power. All these considerations were vital in the competitive tug business.

Reflecting the growing size of tows, 'John Bull' owned by William Watkins was one of the crack Thames tugs in the 1850s. (National Maritime Museum — Greenwich).

By 1849 three more new tugs of increasing size and power were in service, and the Watkins boats were the crack tugs of the Thames. One of the biggest was the *John Bull*, 60 hp, 96 ft overall, and 114 tons - more than four times the size of the old *Monarch*, and with three times her original power. *John Bull* reflected the growing size of tows, for Watkins, faced with an armada of competitors, had to keep up with the needs of the market. In the previous year the four tugs then in the Watkins fleet were already confronting four belonging to Thomas Petley, eight in the Shipowners'Towing Company, eight more in the Thames Steam Towing Company, and no fewer than nine sailing under the flag of the Caledonian Steam Towing Company. The pressure was such that in 1850, Watkins bought his first second-hand tug. Built in 1844 on the Tyne as the *John Lee*, she had been sold first to the Southampton Dock Company for big-ship handling, and now sailed more light-heartedly as the *Saucy Jack* for Watkins.

By 1851 no fewer than 68 tugs were registered in London, although only seven belonged to Watkins. This was a lot of tugs to be chasing business that was still limited to the one long, narrow 'market-place' of the Thames and its estuary. With their limited range, seaworthiness and reliability, the first London tugs did not work beyond Limehouse. As their operators' confidence and experience grew, the limit became Blackwall, then Gravesend, and finally - by 1850 - the Nore Light, which was the seaward boundary of the port of London. The 100 ft *Britannia*, new in 1852, was the first Watkins tug to work the coast in inshore waters, away from the shelter of the river and its estuary. But in practice tug owners did not yet like to venture into distant, exposed waters, where bad weather or other delays could leave them without enough fuel to get home.

The port of Liverpool, though still far behind London, had been a prodigy of growth for a century. About 1722, Daniel Defoe had said that it was 'One of the wonders of Britain ... what it may grow to in time, I know not'.

Liverpool was flourishing primarily on trade with the West Indies and North America. Its merchants exported coal, salt, and manufactured goods from the booming towns of industrial Lancashire, Cheshire, and Yorkshire, which were on Liverpool's doorstep. They poured capital into importing sugar, tobacco, and raw cotton to Britain, and slaves for the British Caribbean plantations. The Asiento Treaty of 1713 with Spain opened that nation's Caribbean colonies also to British slave ships, which multiplied as a result. In 1790 alone, 138 Liverpool-registered ships representing an investment of £1.04 million returned a profit of £357,000, or nearly 34 per cent. Although the slave trade ended in 1807, new markets were opening up. With the coming of peace with the United States in 1815, trade with North America expanded hugely. The winding-up of the East India Company's monopolies of trade with India and China let Liverpool merchants such as John Swire and George Holt into Far Eastern waters for the first time, as it did the entrepreneurs of London.

Food for Britain's burgeoning industrial towns became a significant import: the 8000 tons of wheat imported through Liverpool in 1810 had grown to 75,000 tons by 1850. There was an ever-growing demand for the older imports, too, such as raw cotton for the northern mills. Imports of 40,000 tons in 1810 rose to 111,000 tons within ten years, and to 360,000 tons in 1850. In each case, forty years had seen an increase of ninefold or better. Exports followed the same upward trend. In the year 1852, Liverpool exported 250,000 tons of coal, 315,000 tons of iron, 8000 tons of manufactured goods, and a thousand million yards of cotton fabric.

Such a volume of traffic demanded the fastest and most efficient communications that inventiveness and money could provide, and by 1838, Liverpool had the best railway links of any British port with its industrial hinterland, financed largely by its merchants. The Liverpool and Manchester Railway was opened in 1830; the Grand Junction Railway to Birmingham in 1837; and the North Union line to Preston in the following year (which also saw Liverpool and London linked by the completion of the London and Birmingham Railway).

Where land met sea there had to be sufficient ships to take over the burden. By the 1820s 162,000 tons of shipping was registered in Liverpool - a figure that was rising faster here than in any other major port. In 1820, 7276 ships totalling just over 805,000 tons used the port of Liverpool. By 1840 this figure had risen to almost 16,000, aggregating nearly 2½ million tons. The average size of the ships concerned had risen by nearly half. Although the great majority were very small, by the 1820s the American passenger ships plying between their east coast ports and Liverpool ran to 500 tons and more.

In the years 1851-71 Liverpool's population grew nearly fivefold, to almost half a million. Other British towns were expanding too, but for its size Liverpool was unique in that its population was largely devoted to distribution, not manufacturing: the two largest groups of employees were the dock workers, and the clerks in the offices of the businesses that handled Liverpool's trade. The city grew in order to serve commerce, and depended on it utterly - much more than London did. Liverpool was a port first and last.

The port of Liverpool was shaped like a slim bottle, with a narrow neck broadening into the sheltered Pool beyond. At the beginning of the eighteenth century ships berthed in the neck of the bottle, where the strong tides of the Mersey estuary outside caused fast currents and a big rise and fall in water levels at the wharfs. The growth of trade and increasing quantities of shipping led to an irresistible demand for safe approaches and better accommodation; and thanks to the enterprise of the Liverpool merchants, capital to cover the cost was readily available.

Approach channels were marked and lit; and it comes as no surprise to find Liverpool among the leaders in the development of wet docks. The first dock in the Pool, known later as the Old Dock and four acres in extent, was opened in 1715; and the second, the Salthouse Dock, in 1753. The King's, Queen's, and George Docks followed before the end of the 1700s. By the second quarter of the nineteenth century development was extremely rapid along both banks of the Mersey. Between 1825 and 1846 the area of docks was almost doubled. The Old Dock was filled in; but other existing docks were enlarged and fitted out with the latest cargo handling equipment. New dock building accelerated too. In 1821 the Princes Dock was opened, and in 1836 the Victoria and Trafalgar Docks. By that year seven new docks had been opened in fifteen years, and Liverpool could offer a total of 110 acres of dock accommodation.

Another boom in dock construction started in 1846, when the Albert Docks were opened to traffic. In 1848 alone, five new docks were opened on the north side of the Pool, and four more were there by 1852. On the south side of the Pool, Birkenhead was developed. Its first two docks, Morpeth and Egerton, were opened in 1847. In all, by 1858, the Mersey had acquired twenty-one new docks.

The forces in play were so powerful that tugs proliferated as nowhere else but London. Steam came to stay on the Mersey in 1816, initially - as so often - in the form of passenger boats, engaged in carrying people between or along the increasingly busy and populous banks of the river. But from the first, steamers were advertised as being also available for towing ships out to sea. In October of that year, the Mersey-built steamer *Princess Charlotte,* a passenger boat, towed the *Harlequin* out of the Mersey. Another Mersey-built passenger steamer used for towing, the *Hero,* arrived on the river in 1821. The 30 hp, 64-ton *Druid* of 1823 may have been Liverpool's first purpose-built tug. The *Eagle* of 1824 certainly fell into this category; although all were probably used as general purpose vessels.

By 1836 the growth of the port had convinced Liverpool's entrepreneurs that the steamboat as a tug had an assured future. The Liverpool Steam Tug Company, launched by local merchants and shipowners, bought four tugs: the Liverpool-built 59-ton *Eleanor,* three years old; the *Ormerod;* and the old *Druid* and *Hero.* Two more tugs, built new at their Birkenhead yards, were added to the fleet in 1837: *Queen* and *Victoria* (111 ft) named in honour of the new monarch. All these boats had the advantage of their own repair yard as well, at Wallasey.

The *Queen* and *Victoria* showed an early departure from the idea of a tug simply as a small steamer adapted to towing ships, in no important way different from others, for they were unusually powerful for their day, developing around 100 hp. The reasons may have been the large size of some ships using the port, compared with (say) London, and the fact that

The first Liverpool Steam Tug Company's tug, 'Liverpool Steam Tug No. 1', acquired in 1836, alongside the transatlantic liner 'President' in later years.

Liverpool tugs were beginning to work out to sea. That did not apply on the Thames, where currently tugs were less powerful. The Liverpool company continued to build big. The *President*, added in 1839, and the *Albert* of 1840, (the year of the Queen's marriage), were also 100 hp tugs, of about 82 tons.

They earned most of their keep towing ships between the Mersey docks and the North West Light. A tow part-way, to the Rock Light, cost £3 or £4 depending on the size of ship towed; and another £7 to £12 from the Rock Light to the North West Light. The *Victoria*, and no doubt other big tugs, would tow a ship of any size to the Skerries, off the north-western tip of Anglesey, for £31 10s. Their exceptional size and power enabled them to undertake deep-sea work - the first, and for some time the only, tugs to do so. As early as 1843, the *Victoria* completed a tow from Shannon in the south-west of Ireland to Liverpool, charging a massive £630 for the job.

A feature of early tug work on the Mersey was aid to inshore lifeboats (all, of course, rowed or sailed) - an illustration, perhaps of the difficulties ships got into trying to enter or leave the Mersey in bad weather, for which Liverpool Bay was notorious. In 1837 the *Eleanor* rescued the passengers and crew of the ship *General Gascoyne*. Just after three emigrant ships left Liverpool for New York in January 1839, a violent storm arose. All three were cast ashore and wrecked, among them the *Lockwoods* with 168 passengers. The *Victoria* towed a lifeboat out to the stranded ships. The lifeboat failed to get near the *Lockwoods*, and had to run for shelter, but *Victoria* managed to get close, anchored and took off some passengers. The next day she returned with another tug and rescued a total of 104 passengers and crew from all three wrecks. The tugs were well rewarded for their action. The company was awarded £400; the *Victoria*'s captain received £100, her mate £10, and each of the crew £5. Such payments were generous: thirty or forty years later the normal maximum weekly wage for a tug skipper was £1 15.0, for a mate £1 10.0, and for a deckhand £1.

In the same year, 1839, the Liverpool Steam Tug Company bought its own lifeboat. Later, in common with other tug firms, it towed out Liverpool Corporation and Mersey Docks and Harbour Board lifeboats. In lighter vein, Liverpool tugs also pioneered summer excursion trips, by which otherwise idle tugs and crews were kept gainfully employed in slack spells. In 1836 a trip round the lightship from Prince's Pier cost 1s 6d.

During Liverpool's boom years more than 40 fleet owners and countless single-tug operators worked on the Mersey. Following the trade, they began to multiply in the late 1840s. In 1848 the Liverpool Steam Tug Company had nine tugs at work, and William Willoughby & Son had six. The Mersey Towing Company was busy too. The New Steam Tug Company - shortlived, like so many - was operating three tugs by 1849. The Mersey Towing Company renamed itself the Mersey Tug Company in 1850, but soon found itself with a competitor of the same name, so changed its name again to the New Mersey Tug Company.

Henry Cruse and William Downham, respectively a boat owner and an engineer, set out to impress the customer with the name of their business: the Hercules Steam Tug Company. It launched one tug on the river in 1852, and another in the following year. The Independence Steam Tug Company, too, had a tug working by 1852.

In the same year the brothers William and Thomas Jolliffe joined in the scramble for business with their five-year-old tug *Lion*. William was a master mariner and passenger steamer owner. The Jolliffes come into our story as the second oldest tug operator, after Watkins in London, of those that eventually became elements of The Alexandra Towing Company, and the first to hail from Liverpool, mother port of Alexandra.

The Jolliffes threw their hat into the ring in time to be involved in three momentous developments that were unfolding on the Mersey, the Thames, and all rivers where tugs worked. Some were critical not only for tug operators, but for the future of all of Britain's big ports, and for the nation as a whole. First, existing custom for tugs, arising from ship movements in port, had never stopped expanding. A business depression in 1846-8 had only briefly slowed the growth of commerce. Now, the upward trend accelerated sharply. Second, the design of ships and tugs changed, as their operators sought the most effective ways of exploiting growing markets. Finally, the tug operators' horizons widened, beyond the river banks that had confined them, until ultimately their arena of activity stretched to the remotest corners of the world. The three were interwoven, a new stage in one stimulating innovation in each of the others: but the trade boom was certainly the primary force for change throughout the industry.

3

Blazer

Expansion and Innovation

In 1849 the Navigation Acts that restricted the carriage of imports largely to British ships were repealed. Foreign ships could come and go with almost complete freedom in Britain's home and colonial ports, so long as their own nations gave British ships reciprocal rights. Not only did British trade expand further as a result; so did British shipping, as retaliatory foreign restrictions were lifted.

Britain's ports grew busier with new waves of emigration. London and Liverpool were already the country's main passenger ports - London principally for the voyage to India, taken by merchants, civil servants, and soldiers; Liverpool for emigrants to North America. Emigrants had sailed from the Mersey for the Americas since the seventeenth century, but traffic boomed in the middle of the nineteenth. On average 70,000 people a year had emigrated from Europe in the years 1835-45; in the following decade the figure leapt to 250,000 a year leaving British ports alone. Twenty thousand or so were Irish, most of them sailing from Liverpool.

In 1852 the first shipments of gold from the recently discovered Australian goldfields arrived in London and Liverpool, setting off a new rush of emigrants. Between 1850 and 1854, 1.4 million people left the country. To cope, London built new sailing ships, in particular Blackwall frigates. Liverpool had no suitable ships to meet the new demand, so a local entrepreneur bought a fleet of American passenger vessels, and ordered new ones in North American shipyards. This was James Baines, with his Black Ball Line. His American-built *Lightning, Champion of the Seas, Donald Mackay,* and *James Baines* were crack ships, some over 2000 tons - much bigger than the average deep-sea sailing ship, which now ran to about 1200 tons.

Another massive, if in this case temporary, boost came to the Thames in 1854 with the outbreak of the Crimean War, Europe's first major conflict for nearly 40 years. Most of the army's needs had to be shipped to the Black Sea. Soldiers and stores had to be shifted by lighter to transports and supply ships, most of which were still under sail, and had to be kept moving in and out of port as quickly as possible. Britain's French allies, too, needed help with moving ships; and not only London tugs took advantage. The Thames became a honeypot for tugs from all around the British coast.

As ever, more traffic brought demand for more dock accommodation. Liverpool gained thirteen new docks totalling 176 acres between 1846 and 1864. Rapid growth caused Liverpool to evolve a centralized port authority under city control at an early date. The Mersey Docks and Harbour Board was founded in 1858 with the brief of improving and extending the city's docks.

Heavier traffic, bigger ships, more urgency in turning them around, and more docks all called for more tugs. In London, sixteen new tugs were launched between 1851 and 1854; on the Mersey, another dozen. In the three years following, the Thames operators alone put a further thirty new tugs into service. Watkins had so much trade that when they bought the *Don* for £410 in 1835, it was worth their while to spend another £1400 reconditioning her. Two years later, Watkins dropped their sailing-ship operation, to concentrate all their resources on towing.

In Liverpool, tug operators continued to proliferate. Among the arrivals of 1854 were the St George's Steam Tug Company, started by H.J. Ward, a ship's chandler; and the United Steam Tug Company. In 1857 came John Strong & Sons; another Mersey Steam Tug Company, with their new 110-ton *Great Conquest*; and another New Steam Tug Company which boasted 'the largest and most powerful tug in the United Kingdom' - the 375-ton *Resolute*. An arrival of greater future significance was John Prendiville. By 1856 he was manager of the first Mersey Steam Tug Company (soon to be renamed the Mersey Original Steam Tug Company). New operators brought increased competition, in which outsiders stood a worse than average chance. John Watkins, eldest son of John Rogers Watkins, brought the tug *John Bull* to Liverpool in 1853 to compete for Mersey business, but was driven out after six years. He returned to the family's Thames-side business, as the Watkins representative at Gravesend.

One trend above all others was making new business for tugs. Competition between sailing-ship owners had already introduced a hitherto unknown factor. This was the need for speed in turning a ship round in port and getting it to its berth or anchorage, unloading it, loading it, getting it out to sea. However, quick turn-round in ports helped little while a merchant ship on its journeys between them was at the mercy of the unpredictable wind. The factor that truly sharpened competition was the sea-going steamship which, being independent of the wind, offered a much more reliable timetable than a sailing ship ever could. Because the steamer made time-saving in ports more important, and because it eventually grew bigger than the sailing ship, shipowners became more likely to use tugs.

A regular steamboat service across the Irish Sea had been launched in 1819, and one across the English Channel in 1821. But long ocean voyages by steamship were rare. They posed two problems. The lesser, because soon solved, was the shortage of fresh water to make steam. For lack of space to carry enough fresh water, condensers had to use sea water as a cooling agent; but the jet condenser worked by mixing the coolant and the steam, and so salt passed into the boiler, where it caused corrosion. Furthermore, the salt deposits reduced the boiler heating area, causing loss of power and increasing fuel consumption. At intervals a steamer on a long haul had to heave to or put into port to 'blow down' its boilers; a process which removed the accumulated salt and other impurities.

The second and more intractable difficulty was the inordinately heavy fuel consumption of the standard low-pressure marine engine, in which one system of hot gases generated by the furnace was expected to heat a single large tank of water. These flue boilers needed large amounts of heat - in other words fuel - to create a modest amount of steam pressure. It was reckoned that ten pounds of coal per horsepower per hour was required to feed even a small, *Comet*-sized engine. On this basis a ship with a 220 hp engine would burn a ton of coal an hour. Unless, therefore, coaling stations were available at frequent intervals, which (away from Europe's Atlantic and Mediterranean ports) they were not, steamers took on the appearance of floating coal bunkers. So much more space had to be given over to coal than to paying freight that the ocean-going cargo steamer was economically impossible.

When steam embarked on the ocean, therefore, it propelled passengers and mail, not cargo. The regularity of steam and the reliability of its service commended themselves to the Post Offices of Britain, the United States, and other countries; and also to passengers. Sailing-ship journey times were unpredictable: people had actually starved to death in becalmed transatlantic emigrant vessels. Sailing ships were vulnerable to wind in another sense, too. Passengers in a ship whose master, in competition with others, was under pressure to cram on as much sail as he dared, were liable to a frightening and even dangerous time. For one reason or another, an average of 894 lives a year were lost in British ships between 1833 and 1835. One passenger in 900 died between 1847 and 1873, when sail still drove most of the world's tonnage.

Mails were carried under a government contract, the value of which underwrote much of the cost of the voyage, and was especially valuable to a steamer operator. Mail contracts were freely handed out, as official encouragement for shipbuilding, and thus for the training of seamen. One success story in particular demonstrates the importance of mail contracts in securing the future of steam - and therefore, indirectly in providing a living for tug operators. Backed by a mail contract worth £29,600 a year, the Scots

The 375-ton 'Resolute' owned by the New Steam Tug Company who boasted it was 'the most powerful tug in the United Kingdom'.

shipowner Arthur Anderson set up a London - Gibraltar steam service in 1837. A year later he founded the Peninsular and Oriental Steam Navigation Company, which in 1840 initiated a London - Alexandria service with the Indian mails. Two years later this was extended to India.

Since it operated on routes where coaling stations could be set up, the P&O was spared the extremes of the fuel problem besetting early steamers. This was luridly illustrated when, in 1838, four steamers crossed the Atlantic in competition, within a few months of one another. They were Isambard Kingdom Brunel's 400 hp *Great Western*, the 700-ton, 250 hp Irish Sea steamer *Sirius*, the Channel steamer *Royal William*, and the 1150-ton *Liverpool*. The *Royal William* was so overloaded with coal that her passengers, if they were so inclined and so blasé, could wash their hands in the sea by leaning over the bulwarks. The *Sirius* used one of Samuel Hall's new surface condensers, which kept the boiler water fresh by separating steam from the sea-water that cooled it, and also improved fuel economy. Brunel had built his *Great Western* big - 1320 tons - on the principle that space for essentials such as coal increased at a greater rate than did consumption of fuel by the more powerful engine needed to propel a large steamer. Theory paid off on this occasion: the *Great Western* not only made a faster crossing than the *Sirius* - 14 days against 18 - but had 200 tons of coal left at the end. In spite of her surface condenser the *Sirius* was forced to burn cabin furniture, spare yards, and even a mast in order to keep steaming. The lesson to be drawn was significant for tug operators: ocean-going steamships were going to get bigger.

The transatlantic route did not at first pay without the help of a government mail contract. A contract was first won for steam on this route in 1840, by the Canadian Samuel Cunard and his British & Canadian Royal Mail Steam Packet Company, whose ships plied between Liverpool and Boston or Halifax, Nova Scotia. Nearly half of the internal space of the first Cunarder, the 1140-ton *Britannia*, was devoted to engine, boiler, and fuel bunkers; nearly three-quarters of her carrying capacity had to be given over to coal. Cunard's mail contract was worth £2300 per round voyage, and it has been calculated that without one, *Britannia* would have made over £2500 less per voyage than a sailing ship of similar size. Clearly the contract made a major difference. Significantly, Cunard found £270,000 of the capital he needed within a few days, from the hard-headed merchants of Liverpool and Glasgow.

The American Collins line, with its massive 3000-ton steamers, competed in the North Atlantic in the 1850s with the backing of a US post office mail contract. At first the Collins ships were faster than the Cunarders, and it became a matter of prestige to own the swiftest steamer. By 1855, Cunard had regained the speed record with their *Persia*, 3300 tons and capable of nearly 14 knots. The price was coal consumption at the rate of 150 tons a day.

Incomparable 'Cutty Sark' on her way to her final resting place at Greenwich with two Watkins tugs, one of them the 'Gondia'. This is a detail from the painting by David Cobb which the Company commissioned in 1983 to mark the 150th Anniversary of the founding of what is now The Alexandra Towing Co. (London) Ltd.

Presented to Britain, 'Cleopatra's Needle' was transported by sea in a hollow iron cylinder which had to be rescued by the Watkin's tug 'Anglia'. A further detail from the painting by David Cobb.

One of the most famous maritime paintings of all time. J.M.W. Turner's 'The Fighting Temeraire' 1839 under tow by the tiny 'Monarch', W. Watkins first tug.

Built in 1859 at South Shields for W.T. Jolliffe, 'Lion' was 160ft registered length, 121 tons and 60 n.h.p.

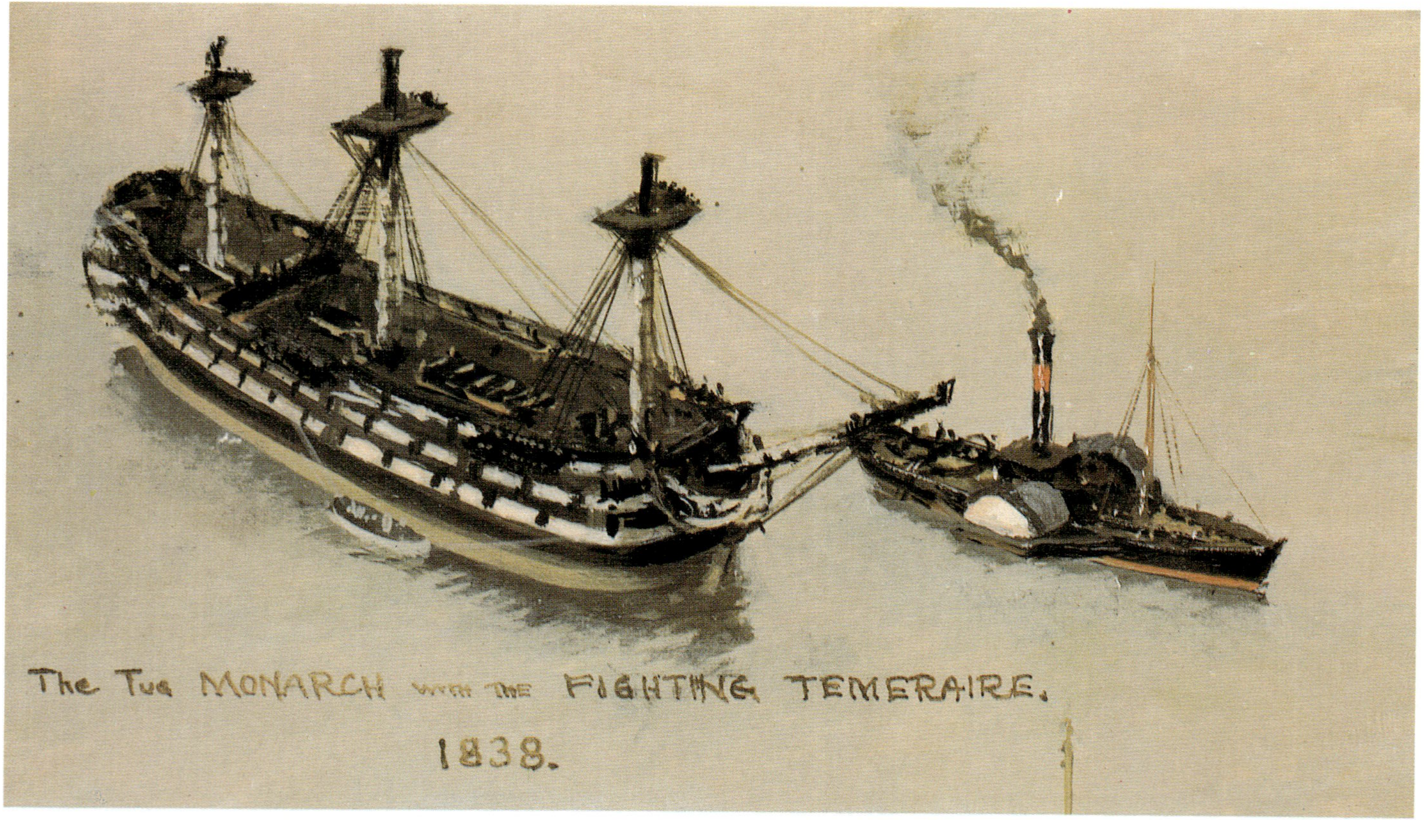

Another artist's view of the 'Monarch' and 'The Fighting Temeraire'. In this case a detail from the painting by David Cobb.

David Cobb's impression of the launch of the 'Great Eastern' at Millwall, the largest ship of her day.

'Tiger' built in 1853 for the Liverpool Steam Tug Company. She was 91ft long, 16ft beam, displaced 107 tons and had 40 n.h.p.

When she was built in 1885, the 975 hp, 149ft 'William Jolliffe' was one of Liverpool's most powerful tugs.

RMS 'Carmania' being towed by Alexandra tugs. She and her sister 'Caronia' were known on Merseyside as 'the pretty sisters'. Painting by Samuel John Milton Brown.

RMS 'Titanic' leaving Belfast in 1912 unladen for her sea trials, from an original painting by E.D. Walker. 'Herald' and 'Wallasey' ahead of her.

Speed and size both demanded tugs. One sign of the success of the big Atlantic passenger ship was the launching in 1848 of the Cunard tender *Satellite*. This new type of boat was basically a big, powerful tug, built with passenger and luggage accommodation so that she could ferry passengers to and from ships moored in the Mersey.

But for all its bumptiousness, the sea-going steamship was still a junior partner to sail in the 1850s. In 1827, a mere 232 steamboats, totalling 23,000 tons, were registered in Britain; in 1837, 531 aggregating 51,000 tons; and in 1847, 924 steamers added up to 116,000 tons - compared with nearly 2.9 million registered tons under sail. As late as this, less than four per cent of Britain's registered tonnage was steam powered; the tonnage of new sailing ships built that year alone almost equalled the total tonnage of British steamships afloat. Three years later the proportion had barely altered. Steam claimed about 188,000 tons, but there was now 4 million tons under sail. The 1847 figures show that, like the average sailing ship, the average steamship was small: 125 tons. This was because most were still confined to rivers, coasts, and narrow seas. In 1835 Lloyd's Register listed only eleven steamers as insured. Clearly, most steamers were not yet expected to venture into dangerous waters.

The amount of shipping under sail did not, at first, suffer from the competition of steam. In fact it expanded greatly; partly because there was usually enough new business for everyone, and partly because steam had an advantage only in the passenger trade. The rapid growth in numbers and tonnage of all kind of ships stands out from the figures. In just three years, 1847-50, we see tonnage increase by 40 per cent. The economics of sail were in its favour. It was not merely that a sailing ship had proportionately more cargo space, thanks to the absence of engines and boilers. Steamships were more expensive to build: the first ocean-going steamers cost £45 - £50 a ton, against £15 - £25 a ton for a big sailing ship. With their superior regularity, two steamers could do the work of four sailing ships in the same time; but a sailing-ship operator could usually get his investment back faster. Furthermore, given a favourable wind a sailing ship was not necessarily slower than an early steamer.

Regular sailings, although not, of course, scheduled arrival dates, had originated in the eighteenth century with Britain's transatlantic mail 'packets', so called because they carried the mails as well as general cargo. Other merchant ships waited until their holds were full before sailing. The packets were forbidden to do so: regularity of sailing was a condition of the mail contract. This made the packets popular with passengers, and 'packet' became synonymous with 'passenger ship'. The idea of the 'liner' also originated in the days of sail. It is a ship that is part of a fleet, its members working together to run a service on a fixed route, sailing at fixed times. The first liners were ships of the American Black Ball line (not to be confused with the Liverpool firm of the same name) that began working across the Atlantic with mail and passengers in 1816. The regularity of their departures was helped by the fact that, by that date, there were steam tugs in the port of New York that towed them clear of land to find a wind.

With the arrival of steam on the open ocean, owners of sailing ships tried hard to win back their passengers in new ways. In the 1850s Liverpool's Black Ball and White Star lines introduced standards of comfort, service, and courtesy hitherto unknown under sail. And since, in favourable conditions, a sailing ship could be as fast or faster than an early transatlantic steamer, it could make an appeal to passengers' desire for speed as well; although comfort would then go overboard. In 1848, Cunarders could reach 10 knots, while an American passenger sailing-ship owner felt bound to apologise for the 11-knot speed of one of his fleet.

On the Australian run, where P&O steamers were working by the 1850s, there was no contest. Their need for frequent coaling stations forced steamships to hug the shores of the Indian Ocean. From Suez they hopped to Ceylon and Singapore before turning south; or else they sailed south-east to Mauritius and then east, with sacks of coal stacked on deck under the passenger's feet. In contrast, once a sailing ship was in the South Atlantic it could take advantage of the prevailing westerly winds to run straight across the Indian Ocean, 8000 miles due east to Melbourne. To return to Britain, it stayed with the westerlies to Cape Horn. Liverpool's Black Ball liners won the Post Office mail contract by guaranteeing to reach Australia in 68 days or forfeit £100 a day.

A generation before steam put in an appearance on the high seas, sailing ships had been improving their performance in competition with one another, growing bigger at the same time. America's Baltimore schooner evolved into the clipper, with its union of size, speed, and economy - the first attribute leading to the second and third, and in no way detracting from the ship's elegance of line. The size of American ocean-going sailing ships grew from around 500 tons in 1820 to 1200 tons in 1840 and 2000 tons in the 1850s. They were cheaper to build than their British competitors, as well as better officered and run. Nor could the typical British Atlantic trader, broader in the beam and bluffer in the bow, though stronger and more reliable, compete in terms of speed. Partly to blame was the 1773 Tonnage Act, which was not fully superseded until 1855. It discouraged the qualities that made a ship fast, for its system of calculating cargo carrying capacity, adopted by ports when fixing the dues they charged visiting ships, saved money for operators of deep-built, boxy vessels.

Then, from the 1850s, British sail underwent a transformation. Not only were more sailing ships than ever before being built; new ships were produced that were the equals of the Americans in their sailing qualities, while retaining their advantages of strength and reliability. Among the pioneers were *Challenger,* launched at Blackwall in 1850, a straightforward copy of an American clipper, and *Lord of the Isles,* Clyde-built in 1856. The new generation culminated in the tea clippers of the 1860s, built to run home the first of the new season's crop from China. They were small - around 800 tons - but they were fast enough to beat the American clippers. The earliest cargo to arrive on the London market - even by a matter of hours - got the best price. The ship' captain and crew were rewarded accordingly, and the clippers made a race of it.

More and bigger steamers, more and bigger sailing ships, all meant more business for tugs - but they also meant that tugs had to change with the trends, if they were to get the custom.

In some ways, changes in tug design mirrored changes in their tows. In 1786 the ironmaster Henry Cort wrought plates for a steam engine boiler from rolled iron. This process made the iron boat possible. The first was a canal barge, built by another ironmaster, Henry Wilkinson. An iron vessel was 25 per cent lighter than a wooden boat of the same volume: cargo carrying capacity could be increased without incurring a weight penalty. Iron construction was stronger than wood, so ships could be built larger. Building in iron could be cheaper; and the new material was longer-lasting.

But wood worked well enough, and it was an entrenched technology, particularly in sailing-ship construction. Not surprisingly, the first iron ships were steamers. Steam engineers, themselves the representatives of one new technology, were perhaps more likely to take a pragmatic view of another. The pioneer was the iron river steamer *Aaron Manby* of 1822. When John Laird of Birkenhead built his first craft in 1829 - a lighter - it was of iron construction, and he was building iron ships by 1834. In 1835, William Fairbairn opened a yard for building iron ships on the Thames at Millwall; six years after that, Robert Napier followed suit on the Clyde. Sea-going iron ships were still rare. Lloyd's insured their first iron ship in 1837, but would not yet do so for a term of years - cover was by the year only. There was a growing handful of iron sailing ships, which proved 10 to 15 per cent cheaper to operate than wooden ones, and soon showed their other virtues. The Laird-built *Garry Owen* was stranded on her maiden voyage in 1834, but was recovered with her hull undamaged.

It was during the 1840s that the iron steamer gained ground significantly. The trend-setting Peninsular and Oriental Steam Navigation Company were early converts. They had their first iron steamer by 1843, and another was on their stocks a year later. More portentous was Isambard Kingdom Brunel's new steamer *Great Britain,* launched in 1845. The lightness and strength of her iron shell and frame made her more economical to run than a wooden steamer. She could be built huge, by the standards of the day - 3270 tons,

322 ft in length, 51 ft in the beam. Cast ashore on a rocky Irish coast for eleven months, she was salvaged essentially unharmed. The lesson had been learned. By 1860, four-fifths of Britain's new steam tonnage was being built in iron. More important, for all new tonnage the figure was over a third, showing (at a time when sail still vastly outweighed steam) that sail was taking to iron hulls as well.

The size of the *Great Britain* called for unprecedented power to propel her. She had boilers of conventional, flue-type design and low pressure, so her engines had to be big to produce the power needed. She had two, each of 500 hp, giving her a speed of 12 knots. The strength of her iron construction enabled her to take advantage of a new means of propulsion - the screw propeller, whose vibration was too much for wooden hulls. A paddle-wheel, most of which was out of the water while turning, wasted a lot of the power generated by the engine. A submerged screw was more efficient; and because it did the same work for the expenditure of less power (and therefore fuel), it was more economical. The Swede John Ericsson had demonstrated a screw-propelled vessel, the *Francis B. Ogden*, on the Thames in 1837. She towed the Lords of the Admiralty in a barge past the Houses of Parliament at nearly nine knots. A year later Ericsson built his screw steamer *Robert F. Stockton*, but after showing her towing four laden coal barges on the Thames, both inventor and boat went off to America, having found too little encouragement in Britain. It was left to the imaginative genius of Brunel to show what a combination of iron and screw could achieve.

The *Great Britain*'s economy of running meant that she had none of the coaling problems of other steamers of her time. She could carry enough fuel in her bunkers for at least twenty-four days. Unlike the P&O liners, therefore, she could run from the Cape of Good Hope to Melbourne direct, as did the sailing ships. Indeed, by the late 1850s, the *Great Britain* was putting up outward bound times of under 60 days, when the crack Black Ball sailing ship *Lightning* might take ten days longer.

The *Great Britain* - and other large steamers too - threw another shadow over the future of sail. Generally, the bigger the ship, sail or steam, the faster and more economical she was. A big ship was also safer for her cargo and passengers, and more comfortable. Yet the size of a sailing ship was limited by the amount of canvas that was manageable; and furthermore, the power of steam engines could be boosted far higher than the energy generated in any sails. Steam therefore allowed much bigger ships to be built. The largest sailing ship of the days of sail was the 10,730-ton *La France*, with more than an acre of canvas. She was launched in 1914, when transatlantic liners had already reached five times her size. As it turned out, the Antipodean passenger market was not big enough for more *Great Britains*, but in the Atlantic, sail's battle for passengers and mail was finally lost. The big, screw-propelled iron steamers of the Inman line out of Liverpool had proved that the new breed of passenger and mail steamer did not need a mail contract to make a profit. They had even appropriated a large slice of the emigrant trade - the bottom end of the passenger market, and the transatlantic sailing packet's last resort.

Iron construction was quickly adopted by tug operators, for its benefits were even clearer to them than to ship owners. As tugs were called upon to handle ever larger ships, they became more liable to damage in collision, so added strength without a weight penalty was welcome. So was added carrying capacity: Watkins's 115-ton iron tug *Punch* of the 1850s could carry thirty tons of coal in her bunkers, which was proportionately a great improvement on the *Monarch*. Longevity was useful in a trade whose tools were worked to the limit, and in which money for repairs and replacements was perennially short. The benefit of cheapness in a business operating on a financial knife-edge was self-evident.

Liverpool led the way, as with so many other advances in tug design. The first iron tug may have been the *Defiance* of 1841; but she was ahead of her time. Seven years later the *Liver* was launched, and in the following decade iron tugs multiplied. Between 1851 and 1857 the Liverpool Steam Tug Company put eleven new tugs in service, all of them iron. New businesses took to iron as well. As its name suggests, the first of the St George's Steam Tug Company' 'King' fleet, the *Iron King* of 1853, was an iron boat. The Jolliffe *Lioness* of 1854 was iron, and also John Prendiville's *Toward Castle* in 1856. By 1854, about twenty-five iron tugs had been built countrywide. Their strength and durability were demonstrated by the fact that four times as many wooden as iron tugs became total losses during the 1880s, even though, by 1885, there were seven metal tugs to every ten in wood.

The screw propeller caught on much more slowly among tug owners. Its power and economy showed best at steady speeds, as when running free or during sustained towing ahead. Its advantages were not very apparent in ship handling, nearly all done in the confined waters of a dock, or on the approaches to riverside wharfs, where the emphasis was on short bursts of power and a high degree of manoeuvrability. In its ability to turn on a sixpence, stop almost dead by reversing its paddle-wheels, go rapidly into reverse, and exert nearly as much power going astern as ahead, the paddle tug was never, in fact, bettered by the conventional screw tug. In particular, the paddler shone in the delicate work of manoeuvring a ship from astern, where the tug acts as the tow's 'rudder'. Stability, too, is vital in a tug; and here the paddler scored. With its broad beam, it was less easily pulled over - girted - by one of its increasingly unmanageable tows.

Furthermore, in their early days screw tugs tended towards shallow draught and small propellers. This made them unpopular with ships' captains, who believed that they lacked grip on the water when towing against the wind. A screw tug, unlike a paddler, could keep her engines

'Blazer', 1888, the second of that name owned by The Liverpool Steam Tug Company — 130 ft, 283 tons, 130 n.h.p. (National Maritime Museum — Greenwich).

well forward so as to bring her towing hook near her longitudinal centre of gravity. But early screw tugs tried to retain the slim, elegant bow lines of the paddler; and the combination of this with the shift of weight forward made them 'dive' when running into waves head on, throwing the screw clear of the water and robbing them of all traction. The continuing popularity of the paddler for some jobs was illustrated by the fact that the Admiralty was still ordering new ones in the 1950s, for manoeuvring aircraft carriers in harbour. The Tees Towing Company's paddle tug *Lingdale* was pensioned off only in 1954, after seventy-two years' service.

With a few of its boats already engaging in long-distance, open-sea work in the 1840s, Liverpool was a natural place to be experimenting with screw tugs. By 1844 the little 65 ft *Liverpool Screw* was in use. But the fate of the Liverpool Steam Tug Company's *Alpha* rings more true: she was found to be useless in the Mersey, and was quickly sold. The first screw tug to go into regular service on the Thames was the *Magnet*, which in 1854 began working with small craft. The screw tug would come into its own; but its time was not yet.

Other changes in the tug were much more noticeable. The slower a ship is moving, the less it responds to its own steering; the bigger the ship, the harder it is for a tug to manage; and once the initial inertia of a big ship at rest has been overcome, its momentum is hard to check. By the 1850s such loads and stresses were helping to make the ship-handling tug a more specialized instrument than it had been in its earliest days. It was getting more powerful, and it was getting bigger. The process had already begun in Liverpool, in response to the challenge of open-sea towing; now it spread to all major ports. The trend was not only necessary; it could also make for economies, for a powerful tug able, on occasion, to do the work of two or more weaker brethren cost its owner less to run.

The Liverpool Steam Tug Company's *Blazer* of 1856 was a monster of 383 tons. Watkins's *Victoria* of 1853 (not to be confused with the earlier, Liverpool *Victoria*) was big - 152 tons - and developed a claimed 450 hp. Boiler pressures had begun to creep up with the general spread of fire-tube boilers in the 1850s; and boiler pressure made power. Prendiville's *Toward Castle* produced it at 40 lb per square inch - then the highest pressure in the British mercantile marine. None of the new power seems excessive when one remembers some of the launches that tugs now had to cope with: Brunel's swan song, the gigantic *Great Eastern*, 680 ft long and nearly 19,000 tons, that finally reached the water in 1859; and Britain's first ironclad battleship, the 380 ft, 9200-ton *Warrior*, launched in the following year.

With higher pressures, engines were becoming more powerful without growing correspondingly bigger. When boiler pressures were low, cylinder size had spelled power. Since an engine's nominal horsepower was based on its cylinder size, its nhp had been more or less the same as its indicated horsepower - the power actually produced at the drive shaft. Now, as pressures climbed, indicated horsepower accelerated dramatically ahead of nominal. This was illustrated strikingly by the New Steam Company's tug *Resolute* of 1857, whose 120 nhp compared with a claimed ihp of 700 - nearly six times as great. But tug owners notoriously exaggerated the actual power of their boats, in order to impress customers. By the 1860s the more usual multiple was four or five; though it went creeping up until the marine steam engine was fully developed. All the horsepowers given in the present context are, as nearly as can be ascertained, those produced at the shaft.

At the same time, the smaller, lower-powered tug retained its role, in work such as towing lighters and the smaller sailing ships such as brigs and schooners, where high power would be wasteful. Old tugs were kept for such jobs, and others added to the fleets. For example, at the time when Watkins were building big tugs, they also bought the 45-ton, 72 ft *Don*. The liner tender had already arrived in Liverpool; as the work done by tugs diversified, the tugs followed suit.

Subtler aspects of design began to change, too, in response to the new needs. A tug's longitudinal centre of gravity, which governed its fore and aft stability, had been unimportant when little power was needed to move a ship. Now, the longitudinal distribution of a tug's parts - their balance - began to loom large. For instance, the position of the towing hook was becoming critical. Tows and turns at full power were impossible if the towing hooks were too far astern of the longitudinal centre of gravity. The hook crept forward, towards the midships point.

Increasingly, tug handling became a skilled business; a matter of balancing not just the familiar drag of wind or tide on tug and tow, but also the trundling to and fro of an ever more massive captive burden. The tug master's latitude for miscalculation, and his time for making critical decisions, decreased in proportion.

Soon after mid-century, the tug's size and power and the developing skills of its handlers were completing its evolution from a creature of quiet waters to one that was at home on the open ocean.

Early Liverpool tugs had shown the way out to the more risky, but lucrative and untapped business beyond a port's confines, but although small tugs did open-sea duty when no others were available and their owners needed the money, the big, powerful tug that spread in the 1850s, often iron-built and with more range than its predecessors, was better suited to it.

One such task was the laying of submarine cables for the electric telegraph, which was extending its filaments beyond mainland Britain. In 1850 tugs helped to lay the link across the English Channel, and the Liverpool Steam Tug Company's *Dreadnought* laid cable between England and Ireland. The

'Uncle Sam'. A Watkins tug from the 1860s. (National Maritime Museum — Greenwich).

219-ton iron Watkins tug *Victor* was chartered for £18 a day in 1861 to repair the cross-Channel cable. Laying and maintaining cables was sought-after work, for it was long-term, avoiding those idle periods that made tug operation so expensive.

Salvage was another task that sent tugs to sea, usually in heavy weather. As early as 1836 the *Hero* - no ocean-going tug - salved the *Columbus* packet: the first recorded salvage in Britain of a ship and its cargo by a tug. By 1845 her owners, the Liverpool Steam Tug Company, were fully equipped for inshore salvage work, with big tugs, submersible lifting vessels called 'camels', and flat-bottomed boats from which divers operated. Among its jobs were the salvage of the *Great Britain* when she ran ashore in Dundrum Bay in 1846, and of the iron sailing ship *Tayleur,* also wrecked on the Irish coast in 1854. The 336 ft, 2719-ton iron screw ship *Royal Charter* was completed in 1855 as a running mate for the *Great Britain,* but ran aground in Moelfre Bay, Anglesey, in the great gale of October 1859. She disintegrated in seas no ship could survive, with the loss of 454 lives; but the boats of the Liverpool Steam Tug Company and the United Steam Tug Company salved most of her half-million-pound cargo of gold. At around this time on the Thames, Watkins tugs also began to find salvage work. In 1859 the *Victor* and *Victoria* rescued the 1700-ton iron sailing ship *Ailsa,* for which they were awarded £600. Two years later the same tugs, aided by another Watkins boat, the *Uncle Sam,* salved the sailing ship *Aithona,* and this time netted a record £1250 between them.

Throughout the nineteenth century big salvage claims were settled by the Admiralty Court, before which representatives of salvors and salved routinely put their case. Smaller claims went before the County Courts. Awards were based on the value of the ship and its cargo; on the amount of work and danger involved for the tug, taking into account the prevailing weather; and the share of responsibility for the salvage that the Court felt a tug deserved. After an award was made, a tug operator shared out a proportion of it among the crew. Salvage work could be very lucrative - three small second-hand tugs could have been bought with the *Aithona*'s salvage award alone - but its rewards were correspondingly hard to win, and sometimes were altogether denied. The tug owner had to fight for his money. The ship's owner would do his best to minimize the part played by a tug in a rescue, claiming that his own master had made a significant contribution, or even could have coped on his own, without the unwarranted intrusion of the tug. If he could, the ship owner also minimized the value of his ship or its cargo, or both, particularly if either had been damaged. A tug sometimes had to share an award not only with tugs of the same company, but with those of any other operator who could show that he took part in the salvage. The value of that part would be contested too; an aspect of the perennial rivalry between tugmen. If it was unavoidable, a salvage was shared from the outset between different operators, and the proceeds split. In 1877, Watkins' *Victoria* and *Renown* got £1600 of an award, and a rival tug £800.

Port salvage associations were organized by local Lloyd's marine insurance underwriters and owners of ships and cargoes, as a means of minimizing their losses. They made agreements with reputable owners, in order to ensure efficient, reliable salvage at economic rates. Such contracts were to the advantage of the tug owners, too, for they were a source of steady income - the elusive ideal of all operators. Watkins and other owners chartered tugs to salvage associations, putting them on standby for perhaps weeks at a time for a fixed fee per tide. In 1863 £25 per tide was asked for the *Victoria*, and £20 for other Watkins tugs.

Open-sea towing spread from Liverpool to the Thames in the mid-1850's. In 1855, Watkins's *Victoria* and *Britannia* undertook their first tows to Calais, charging £105 for a 1070-ton iron sailing ship. In the same year the *Victoria* towed a ship to the Texel in Holland for £325. Two years later Watkins built the 157-ton *Napoleon* specifically for long haul work. In 1860 she went from the Thames to Plymouth to pick up a sailing ship, then towed it to Flushing in Belgium: a total of about 800 miles. The *Napoleon* was equipped with a steam-powered windlass for rapid weighing of the anchor and for towing-off grounded ships; but she was conservative in retaining a wooden hull,

The twin-funnelled 'Renown' in 1879. 70.1 ft, 54 tons, 22 n.h.p. (National Maritime Museum — Greenwich).

for which her engines proved too powerful. On journeys beyond the capacity of a tug's bunkers coaling stops were necessary. The *Victor* burned 104 tons in 132 hours' running from London to Liverpool with one tow and back to Swansea with another - a consumption rate of 19 tons per day for 5½ days.

Bigger ships called for deeper channels in ports worldwide, which often lacked dredgers. These vessels had steam engines to work their dredging machinery, but they were not self-propelled, and had to be towed into place. In good weather this was a straightforward task for the new breed of tug - in 1860 the *Victor* towed the *Navigator* from Gravesend to Cadiz for £75 10s - but a dredger was an unseaworthy craft as well as a dead weight, and winter weather, especially in the Bay of Biscay, dramatically increased a tug's difficulties. Watkins charged £786 to tow the *Industrious* to Ferrol in Spain in November and December 1860.

The new tugs were able to do more than accept specific long tows. The same qualities enabled them to go out looking for business 'on spec'; or to put it in the more elegant and romantic terminology of the day, they could go 'seeking'. More sailing ships might now be departing at fixed times, but their operators could never honestly promise punctual arrival, as could owners of steamers. When, for instance, a sailing ship approaching the Thames encountered westerly winds in the Channel, in the Downs, or off the North Foreland, she might have to anchor for days or even weeks, but if her master could expect to meet a tug there, she was home and dry. And tugs aided not only the ship held up or becalmed, but also the damaged, disabled ship whose captain would be pleased to reach any port, whether or not it was his intended destination: the job was then salvage, and much more lucrative. An unscrupulous tugmaster was not above 'improving' a situation. In bad weather, when visibility was poor, he might let a tow slip, allow it to drift towards shore, pick it up again, and claim salvage; but the Admiralty Court was wise to this practice. Although a big-ship tow commanded the highest charges, if none was to be had a tugmaster did not disdain to seek smaller fry. These included becalmed trawlers anxious to get home with their catch, which could be towed to port in shoals; and the lowly Baltic firewood carriers called 'onkers'.

London-bound ships might be delayed further from home than the Downs, and as tugs' endurance grew, so they worked further afield to pick up trade. Beachy Head, the Isle of Wight, the Scillies, and eventually - by the time sail had finally lost its battle with steam - the south-west coast of Ireland became the limits of seeking for Thames tugs. As they worked their way west, so tugs from south coast ports such as Southampton would join in the hunt; then boats from Plymouth, Falmouth, and the Bristol Channel ports. At the Fastnet light off the south-west of Ireland they would encounter Liverpool and Glasgow tugs from the north, seeking down the Irish Sea - Prendiville, Jolliffe, the Liverpool Steam Tug Company. Many Liverpool tugs towed sailing ships out to the Tuskar Light, in St George's Channel off the south-eastern corner of Ireland; then sought an incoming sail to make the return journey pay. Meanwhile, off the east coast, tugs out of the Tyne, Tees, and Humber cruised the North Sea for business.

The idea of seeking, as far as the Thames was concerned, was the brainchild of Richard Ross of Lloyd's Steam Tug Company of Blackwall, who in the early 1850s was building up yet another tug fleet on the river. He was sure that tugs could now be built to work beyond the existing accepted limit of the Nore, down the Channel as far as the Downs, towing ships out to that point and picking up incoming trade. Ross planned big tugs to do this, but was anticipated by Watkins, who in 1853 launched their *Victoria*.

Built by Money Wigram of Blackwall - constructors of the Blackwall frigates - she was of strong, composite wood and metal construction. For economy of operation she had two boilers. One was for ordinary running; the other would be fired when a sailing ship was reported and extra speed was needed. Economies were essential in seeking. It was the most wasteful form of tug operation, in which a boat might not only use quantities of fuel to no purpose, but might also run out of it at a critical moment. Even so, the bunker capacity of the *Victoria* was a modest twenty tons. There was a steam windlass, as on the later *Napoleon*. She was the first Thames tug built specifically for seeking; though she was used equally for any other long-distance, open-sea work.

In 1866 the Thames 'seeking' paddle tug reached its apotheosis in the Watkins boat *Anglia*, a fine sea boat. She had three fire-tube boilers producing 700 hp - enough to handle the biggest tows in the world - and three funnels to match (one forward, and two abreast behind it) that gave her the nickname 'Three-Fingered Jack'. But she also needed the biggest bunker space of any tug afloat - ninety tons - for her appetite for coal was awesome. On her trials, thirty tons went up in smoke between Gravesend and Beachy Head.

When a seeking tug left port she normally carried coal and food for seven days, after which she would have to put into port to re-stock. Her deck, as well as her bunkers, were stacked with coal, to give her the greatest possible range without refuelling, a practice that could make her dangerously top-heavy. A big sea would sometimes carry the deck cargo overboard, together with a length of the bulwarks - if the crew, afraid for their lives, had not already ditched the coal and rearranged the bulwarks with an axe to make it look like an accident.

If possible, the tug going seeking made both legs of her trip pay by towing a ship out with her, dropping the tow at sea, and then looking for an incoming sail. The tugmaster - or rather, his engineer - cruised as thriftily as he knew how, using the sails as much as possible. The master ordered full speed only when he saw the topsails of a potential customer, or to get

'Anglia', 1866 William Watkins 'seeking' tug. Three fire-tube boilers producing 700 hp and boasting three funnels: hence her nickname 'Three Fingered Jack' — 144.4 ft, 275 tons. (National Maritime Museum — Greenwich).

ahead of another tug on the same errand. He came by his knowledge of when and where to look for a ship in a variety of ways. Some was stored in his head before he sailed. He studied the *Shipping Gazette, Lloyd's List,* or the *Liverpool Journal of Commerce* for news brought by steamers of approaching sail. Adding to this his own familiarity with the intimate characteristics and past performance of ships, captains, and owners, and bearing in mind wind and weather, he tried to work out likely positions and probable arrival times. The end of a spell of favourable wind in, for example, the Downs, was a good time for tugs, for a lot of ships would tend to arrive together, pushing up demand for a tow and hence its price.

At sea, the tugmaster 'spoke' any passing steamer he met, to ask for news of sailing ships that the steamer had overtaken. He also kept his eyes open when passing a port or fishing village. Tug-owners had watchful agents ashore who telegraphed with news of ships, typically local boatmen or publicans, who had good contacts among ships' crews and pumped them for information and gossip. When an agent had news for a tugmaster, he ran up the company's house flag at some prominent vantage point. This was the master's cue to send a boat ashore for a briefing. Alternatively, the agent and master exchanged coded signals by flag or semaphore, which agents of rival tug operators tried to decipher.

The rival tug was another element in the equation. A tugmaster should try to get his boat lower down the Channel, or the Irish Sea, than any rival; but he must not waste fuel by going too far. During a spell of fair winds he must anchor, in order not to waste fuel in fruitless seeking. On at least one occasion Charles Gaselee's tug *Revenge* lay off Beachy Head for weeks before finding a tow back to the Thames.

Once he received information about a ship, the tugmaster had to keep his movements secret from other tugs with whom he might be congregating, in favoured spots such as the Downs, Beachy Head, or off the Isle of Wight. He preferred to weigh anchor before dawn, with silenced windlass, moving off without navigation lights, or with lights reversed to give the appearance of going in the opposite direction. In his hunt for the 2½ per cent commission on the price of a tow that he could usually expect from his owner, any trick, however illegal, devious, or dangerous, was considered fair, up to the moment of contact with a potential customer. Once when the Watkins tug *Manila* went seeking with lights doused, her master forgot to display them again when he hailed a ship. Its captain enquired crossly where they were, upon which the tugmaster hazarded the guess that the tug's cat had eaten them. He got his tow.

For his part, the sailing-ship master in imminent need of a tow scanned the horizon for a smudge of smoke. While the tugmaster always had a free hand to get the best bargain he could from a ship, his opposite number might not be so lucky. He must not, in any case, accept a tow until he had lost the last puff of fair wind, for the closer he was to home, the greater the number of competing tugs he was likely to meet and the shorter the tow. Both factors reduced the price he must pay. In turn this increased his commission on the net earnings of the voyage, and pleased his owners. On the other hand, in the atmosphere of competition now growing even in cargo carrying, there might be pressure on him to make port as soon as he could; and certainly he, his crew, and his passengers would be eager to see shore again, after perhaps months at sea.

The climax of the drama came when two or more tugs from different companies sighted a 'prospect'. Shrugging off the possibility of a boiler explosion, the engineer might tie down his safety valve to extract the last knot of speed from an already highly-stressed engine. The boiler was the most vulnerable and most often replaced part of a tug, its rivets popping and weeping under alternate heavy strains and comparative idleness. When there was stiff competition for business, tugs had to be kept in service somehow, and were much patched to keep them afloat.

In the Channel, an unwritten rule said that the first tug to hail a ship - 'Are you going to take steam, captain?' - had the right to negotiate first for the tow. On the north-east coast, custom laid down that once a tug drew more than a hundred yards ahead of a rival, it won priority. In practice, the prize usually went to the fastest; and even then might be contested. The tug *Nimrod* out of Sunderland, already alongside a ship, was rammed clear by the *Marco Polo*, her master claiming that he had made first contact. In 1866 two tugs out of South Shields raced for a barque off the Tyne. The *Adur* was ahead but the *Scottish Maid* was faster, so that they came up to the ship together. The *Scottish Maid* ran across the bows of the *Adur*, her captain pelting his rival with a bottle and two-inch iron nuts as he did so. Since, at the time, there were no fewer than 236 tugs and passenger boats working on the Tyne, the bitterness of competition is easy to understand. On being refused by a ship, one tugmaster deliberately rammed her in his frustration.

Once a tug had drawn alongside, and the ship's captain had decided to take a tow, the two masters would negotiate a price on the move. An early attempt to fix a tariff of charges, based on distance, for tugs working out of the Thames, was short-lived - there were too many variables. There was only one constant: if possible the tow provided the towrope, for if the tug did so, there was an extra charge. Otherwise only haggling was normal.

Tugs held a strong hand when time was critical for a ship, as when a China tea clipper with the first of the new season's crop ran out of wind, and her captain saw his bonus disappearing with it. John Masefield gave his Captain 'Cruiser' Trewsbury of the tea clipper *Bird of Dawning* an experience which he clearly regarded as typical. Worried about how close his rivals might be behind him, Trewsbury telegraphs for tugs of the 'London and Dover Tug

Company' to meet him. But before they do so, 'two pirates' of the 'South Foreland Tug Company' come alongside. They demand £100 a tug for a tow to the Thames. Pirates or no, Trewsbury - without a second thought for his prior commitment - offers them £50 each. The 'pirates' react by sheering off in 'the familiar gesture of tugmen driving a bargain', and finally depart to succour a stricken rival of the *Bird of Dawning*. But off Dover the 'tall scarlet smokestacks banded at the top with black' of the London and Dover Company's tugs heave in sight, and our hero goes to win the great tea race. Such tactics were regarded as fair enough, in a cut-throat game. Only the most blatant exploitation was condemned, as when the tug *Triton* out of Falmouth charged the becalmed barque *Medway* £40 for a two-mile tow into port. On another occasion, the Australian clipper *Cimba* met a tug off the Eddystone Light, abreast of Plymouth, and was asked £50 for a tow to the Thames. The ship's captain still had a breath of wind so he offered £30, which the tug refused. But when the clipper reached the North Foreland and was stopped by the westerlies, her master found many other ships clamouring for tugs, and eventually had to pay £60 for what, by then, was a short tow.

Haggling was acceptable even in thoroughly dangerous situations. Returning from dropping a tow, a Liverpool tug found an incoming sailing ship off the Tuskar light. Its captain offered one-third of the tugmaster's price, so he departed to look for a more profitable customer. He failed to find one, and when a gale blew up, sought out his original 'prospect', finding her being blown ashore on to the Welsh coast and now only too glad to offer £250 for a tow. The tugmaster refused again; and only passed the ship a line when her captain agreed to take it at any cost. With his ship safe in Liverpool, her owner offered £350, but on taking him to the Admiralty Court, the tug owner got £650. Nerve, and a strong streak of ruthlessness, had paid off.

The sailing ship *Alexander Yates* had a lurid reputation among tugmasters. Her captain bluffed two of them into a £200 tow before they realised that she was helpless, having lost one anchor, dragged the other, and broken her windlass. The tug company put in a claim for salvage. The shipowners paid £1085 into court, but this was refused, and the tugs eventually got more. Later, the *Alexander Yates* was sheltering below Dungeness when her master refused a tow from the Watkins tug *Nubia*, believing that other tugs would arrive and enable him to negotiate a lower price. But it began to blow hard, compelling him to accept a tow from the *Nubia* in the end.

Bargaining followed a predictable course. The tugmaster pleaded the price of coal, or the terrible luck that had come his way of late. His opposite number countered with harrowing stories of absurdly low freight rates, and owners whose attitude to towing expenses would have shamed a miser. To oil the wheels of negotiation, the tugmaster would offer fresh bread, meat,

An advanced example of a river tug, 'Nubia' built of steel — single screw tug with 350 hp triple expansion engines. Owned by Watkins, she handled P&O liners in the 1890s. (National Maritime Museum — Greenwich).

and vegetables, and recent newspapers for the news-starved crew. Sometime's the ship's crew, anxious to re-acquaint themselves with the beckoning delights of port, offered to chip in with the difference between their captain's best price and that asked by the tug, but a tugmaster was generally reluctant to take money from a ship's crew. He was more likely to accept it from the captain, who stood a chance of being reimbursed by the ship's owners. The final agreement was always verbal; once a price was settled, neither master went back on it.

On a dark night with the wind and sea rising, both captains' troubles now began. These have changed little over the years. Their first task was to get a passing line across from tug to tow. The towrope was fastened to it and passed across: hence its name. The rope or hawser, 90 to 120 fathoms (540 to 720 ft) long for ocean tows, was of manila hemp and about a foot in circumference. From the 1870s it was usually attached to a length of three-inch steel wire rope passed through the tow's hawse holes, for greater strength. But the manila was always vital, to give the rope 'spring', the elasticity that helped to prevent a towrope parting under a strain. The tug's crew passed the rope over the tow-rail, which carried it clear of the deck, and dropped the looped end over the towing hook. In more sheltered waters, where a tug needed less space between it and the tow, and where delicate manoeuvring was likely to be called for, a 30-fathom rope might be used.

Successful ship handling in heavy weather and poor visibility, with two vessels practically on top of one another, was part skill and part a matter of luck. One or the other might fail, as when the barque *Wave Queen* came crashing down on the tug *Ocean*. Funnels dented by plunging bowsprits were routine. The tugmaster's first concern was to prevent the passing line or towrope fouling his paddle wheels or screw, for a tug left suddenly without power in a big sea was a tug helpless, and maybe doomed. If the weather was too bad, the tug would act as a sea anchor for its tow until it relented, to prevent it drifting; or else the tow would anchor, with the tug taking the strain off her anchor cable to stop it dragging.

When a tow began the tug's engines went slow ahead, to take the strain gradually. Sudden tension during a tow - difficult to avoid entirely when both vessels were pitching violently, and keeping station was impossible - could mean a broken towrope, a towing hook torn out by the roots, and a crewman cut in half by the backlash of the severed end. A rope could kill or cripple even when, as the angle of the tug to the tow changed, it whipped a few feet from one side of the tow rail to the other. The tugmaster kept his rope as long as possible, for easier management of the tow in a swell, and to guard against sudden strains. Some ships towed sweetly - 'towed like a boat', in the tugman's phrase: others 'towed like a town'. These were the ships that had bottoms foul with marine encrustations, or had been kept in service until they were literally falling to pieces, and had to be bound together with wire rope. Such floating nightmares were common in the last days of sail, when sailing ships were losing money and there was none to spare for repairs. The drag of any big, high-sided ship in a wind was immense. A tugmaster preferred a laden to an unladen tow, which might be blown about unmanageably. Occasionally, a tug with a big tow, battling against a headwind, might seek help from another, taking care if possible to keep the tow 'in the family'.

At the seaward boundary of a port, such as the Nore light on the Thames, a seeking tug customarily dropped its tow. With a hammer, a deckhand tripped the hook, and the rope slid away over the stern, to be wound in by the ship's windlass. At the Nore - or off Anglesey if she was inward bound for Liverpool - the incoming ship was obliged to take a pilot on board, and if necessary, the London-bound captain engaged one of the smaller river tugs that congregated at the Nore for the final run in to the ship's moorings. Many pilots on the Thames were tugowners as well, working out of Gravesend. Some ports forbade this combination of occupations, for it was obviously open to abuse: it was certainly resented by other tugowners. A pilot had the power to insist on a ship's captain taking a tug; and a pilot who owned a tug was suspected of making that captain take one unnecessarily. Even if they did not own tugs, pilots were sometimes accused by tugmen of taking bribes to favour one tugowner above another. Often a visiting ship's captain did not know enough about a port's tugs to make a rational choice - for example, which one was powerful enough - and had to rely on the pilot's advice. In recognition, the tugmaster chosen gave the pilot a present. If he failed to do so, he could expect that pilot to advise in favour of another tug company when he was next called on.

4

Anglia

Boom Years

Occasional trade depressions punctuated the growing prosperity of Britain's two great ports, without ever coming near to stopping it. In the forty years from 1870, the nation's foreign trade doubled in value and more than doubled in volume; while between 1873 and 1903, exports went up in value by nearly two-thirds. To carry this trade, in 1875 there were 6.1 million tons of shipping registered under the British flag: an increase of 45 per cent over 1850. By 1885, the figure was 7.4 million tons.

London's docks expanded to keep pace. Existing docks were enlarged and deepened to suit bigger ships, and new ones were built. By the time the Surrey Commercial Docks opened in 1860, London had 165 acres of wet dock, spreading from the Pool below the Tower to Limehouse Reach, stretching out towards the sea as ever deeper moorings were called for. The Millwall Dock Company was formed to build 36 acres of accommodation in 1864, and the South West India Dock was opened in 1870.

The Royal Albert Dock got its Act of Parliament in 1875. Meat and butter had been luxuries for Britain's urban working class, but freezer ships brought them within almost everyone's reach. The first cheap frozen beef and lamb came from Australia in 1880, and the Royal Albert Dock came to specialize in such products. In 1882, it saw the first shipment of frozen New Zealand meat. Imports rose from nil in the 1870s to 700,000 tons by 1887. The 16 acres of the Canada Dock came into use in 1876. Finally, in 1886, the Tilbury Docks were opened, twenty miles downstream from London, providing 90 acres of deep water moorings for ocean-going ships.

As ever, more and bigger ships and more docks demanded more tugs. And as before, the wider spread of docks along the river called for more lighters, to carry goods from ships or docks to warehouses upriver. More lighters also meant more work for tugs. Business for Thames tugs declined in only one direction. In 1865 British shipyards for the first time built more steamers than sailing ships. Five years later, five-sixths of new tonnage, steam and sail together, was iron-built. More ships propelled by steam and made of iron meant that shipbuilding - and hence ship launches - began to move away from London to rivers with coal and iron on their doorsteps, such as the Clyde, the Tyne, the Wear, and the Mersey.

Not that the Mersey needed any help from London to flourish. Between 1870 and the First World War, Liverpool was the rising star of Britain's economy. On and off, trade was relatively slack on Merseyside for a dozen years before 1888; but the general trend was upward. The tonnage of ships using the port rose from 4.4 million in 1858 to 7.5 million in 1880, and 9.7 million in 1890. Their numbers revealed a significant trend. The 19,500 ships of 1870 totalled 5.7 million tons; the 23,600 ships of 1890, 9.7 million. Ship numbers had risen by 21 per cent, which was good news for tug owners; but ship size had gone up twice as fast, which was better.

The flow of imports through the Mersey accelerated, trebling in value between 1850 and 1913, and trebling or quadrupling in volume. Cotton was king: through the city came 80 to 90 per cent of Britain's imports of the raw fibre. By the 1880s around 450,000 tons a year was being imported. Although the American Civil War cut off the supply of raw cotton from the southern states, other imports poured in unchecked. In 1852, for instance, 121,000 tons of grain came in through Liverpool, but in the following half-century the annual quantity imported increased almost tenfold. Exports were still more buoyant. In the years 1850 to 1913 they went up fourfold in value and fivefold in volume. For Liverpool, the export of people was bigger business than ever, making the shipowners a profit in those rare spells when trade in goods was bad. Of the 5½ million emigrants who sailed from Britain in the forty years from 1860, all but 750,000 passed through Liverpool.

Boom times justified, and paid for, massive new investments in docks, warehouses, and other port facilities. The first ten years of the Mersey Docks and Harbour Board saw £6.4 million spent on Merseyside. In the thirty years from 1870 to 1900, £74 million was ploughed back into Britain's ports, more than £16 million of it - 22 per cent - into the one port of Liverpool. Between 1881 and 1888 six new docks of bigger average size than hitherto, totalling 123 acres, were added to the city's resources north of the city centre, to cater for larger ships. One of them was the Alexandra Dock, of 44½ acres, opened in 1881 and named after the Princess of Wales. The southern docks towards Garston were rebuilt. At Birkenhead across the river, there were 165 acres of wet dock by 1872.

Much of the capital generated by Liverpool's trade went into the building of iron screw steamers. The trend reflected a nationwide surge in steamship building on an unprecedented scale. In 1875, there was still 4.2 million tons of British shipping under sail, against 1.9 million tons steam-driven; but seven years later steam tonnage exceeded that of sail for the first time, and by 1885 it totalled 4 million tons, compared with sail's 3.4 million tons.

But there was more to the new growth of steam than money. The fact that the volume of Britain's foreign trade was growing faster than its value was a pointer. A larger proportion of high-volume, low-value, basic goods than before - grain, sugar, raw cotton, metal ores, coal - was being exchanged. Coal exports stood at 1.4 million tons in 1840, and 49 million tons in 1887 - a thirty-five-fold increase. The implication was that sea transport costs were lower.

The first, and most crucial, of the developments that made this possible was the compound steam engine. In 1856 the Pacific Steam Navigation Company, founded in 1840 by the American William Wheelwright, but with British capital, had introduced John Elder's invention into their ships on the run between Liverpool and the Pacific coast ports of South America. It was so called because it used the same steam twice, in two cylinders in succession instead of one, driving a piston in each. This more efficient use of steam allowed more power to be extracted from a given amount of coal, thereby making the marine engine dramatically more economical. Not surprisingly, the P&O was the first company to follow suit, installing compound engines in its liner *Mooltan* in 1861; but in 1862 the Liverpool shipowner Alfred Holt improved the design, making it still more frugal, and ensuring its triumph. In the following year three 2280-ton Holt steamers were built on the Clyde for the China trade - cargo steamers, for a part of the world hitherto the preserve of sail. Fully laden with cargo, each could also carry enough coal to steam the 8500 miles from Liverpool to Mauritius without refuelling, at a speed of ten knots. A Holt compound engine of 1863 burned only 2¼ lb of coal per horsepower per hour - less than a quarter of the ordinary steamship's consumption.

Now that the ocean-going cargo steamer was suddenly an economic proposition, other shipping companies were created, or recreated, on its back. Most of the new initiative sprang up in Liverpool from the late 1860s. The brothers Thomas and James Harrison were able to reach out to the Gulf of Mexico, the West Indies, Latin America, and India; Alfred Booth to South America; Elder Dempster to West Africa. No part of the world was closed to the cargo steamer. The advantages of compound engines were hardly less obvious in other trades. The Cunarder *Java* had them by 1865; the Inman line and other transatlantic passenger operators followed. Nearer home, the tugs of Britain's home ports would take to compounding, as well as to other innovations that were improving steamship efficiency.

In a development of the compound engine, the triple expansion engine introduced in 1881, steam was used not twice but three times, in three cylinders. The use of steam superheaters spread from the 1860s. In superheating, the wet steam produced by a boiler was dried before it was admitted to the cylinders, raising its temperature and pressure, so enabling it to perform more efficiently. Improved fuel economy was the outcome.

The new steamers were powerful as well as cheaper to run, thanks to boiler pressures now of the order of 60 lb per square inch. The fire tube boiler, in general use by the 1860s, was just one of the several developments that were encouraging higher pressures. Another means of enlarging the heating surface of a boiler, and hence its efficiency, began to appear from the 1850s - the water tube boiler. In this design the water passed in numerous narrow tubes through the furnace; the principle of the fire tube boiler in reverse.

Still higher pressures, up to about 200 psi, were made available by two other developments. From their beginnings, marine boilers had been square, but this 'box' form militated against high pressures. The 1880s saw the introduction of the cylindrical boiler, a form which equalized steam pressure over its inside surface, so permitting that pressure to be higher. By the 1860s the Bessemer converter and the open-hearth furnace had made possible the mass production of cheap steel. Its quality had improved sufficiently by the 1880s to allow its use in ship and boiler construction. Steel had all the advantages of iron except price, only to a greater degree. It was lighter, it had greater tensile strength - resistance to stretching - and it could be beaten or drawn thinner. The capacity of ships could be increased again, without increasing their weight; and they could be built bigger. And, of course, boilers could be made to withstand greater pressures.

The powerful engine that was also frugal could produce combinations of speed, size and economy unheard-of in the 1850s. A good example was the 5000-ton transatlantic liner *Britannic* of 1874. Her engines developed 5500 hp, propelling her at 16 knots; but her coal consumption was only 75 tons a day.

The most catastrophic failure that a normal single-screw steamer could suffer was a broken propeller shaft. As Joseph Conrad said, 'Of all ships disabled at sea, a steamer who has lost her propeller is the most helpless'. From the 1880s, twin screws began to appear: a ship with one broken shaft could usually limp home on the other. Twin screws also made for economy, for they gave greater speed without a proportionate penalty in extra fuel consumption.

The cumulative effect of these developments on shipping, shipbuilding, and the tug was revolutionary. The most obvious and dramatic aspect was the consequence for sail. Even while the finest flowering of a 4000-year old tradition - the China tea clippers - were being built on the Clyde, the Holt steamers that spelt their doom were under construction in an adjoining shipyard. At the end of the 1860s it took the clippers around ninety days to run home to London with the tea: it took one of Holt's cargo steamers sixty-five days. The fatal blow for the China tea clippers was the opening of the Suez Canal in 1869, which reduced the journey time of steamers by ten days or more, and in so doing cut their costs as well.

However, the tea clippers, though the most prominent of Britain's sailing cargo carriers, were only a tiny proportion of them, and the most vulnerable. The ascendancy of the cargo steamer took more than twenty years to develop, and sail fought back every inch of the way. The clippers fell back on the Australian and New Zealand bulk freight run, taking out coal and bringing back wool and wheat. The less specialized sort of British ocean-going sailing ship continued to be built in large numbers well into the 1870s. The 1860s had seen a last, massive increase in its tonnage. There had been just under 5 million tons registered in 1865, and a record 275,000 tons had been built the year before. A decline had set in by 1875, when the total at sea came to 4.2 million tons, but there was still a greater tonnage of British sail than there had been twenty-five years earlier. Sail shared, to some extent, steam's economic advantages of increasing size and iron construction; and it adapted to the competition of steam with economies such as the mechanization of sail and anchor handling, which reduced crew numbers. Only with the general introduction of the triple expansion engine did sail succumb to the logic of superior economics; only then did tugs find that their original clientele was disappearing.

The decline was gradual and, as far as tugs were concerned, more than compensated for by the increase in steam business. It was not just that there was a whole new type of ship flooding in and out of Britain's ports: there was a dramatic increase in the size of steamers in general. Passenger liners, particularly on the prestigious transatlantic run, were growing even larger. In 1867 the shipowner Thomas Ismay bought the failed White Star Line, whose famous Australian sailing ships had had their day, and took his acquisition into the North Atlantic passenger trade. In so doing he transformed the passenger liner, and rendered all existing ships obsolete. Ismay's philosophy was that creature comforts for the paying customer were the source of profit in this operation, and so must be the operator's overriding concern.

Ismay's 3700-ton *Oceanic* of 1871 and its successors were the first passenger liners of modern design in this, and other, respects. Passengers' cabins were large, light, and airy, and installed amidships, forward of the engines, not towards the stern - their traditional location in sailing ships, but now above the vibrating screw. The navigating position was a bridge above the engines, for ease of communication with the engine room, instead of astern above the rudder. Rails replaced solid bulwarks, to allow heavy seas to drain away. For speed, the *Oceanic* was long - 420 ft - in relation to her 41 ft beam. Her engines delivered 3000 hp, and she had a speed of 14¾ knots. Ismay's *Britannic* of 1874 improved on the formula. She was a 5000-ton ship carrying 1300 passengers; longer, but with the same 10:1 or better ratio of length to beam. Her 16-knot speed made her the fastest ocean steamer afloat. Ismay's competitors - Inman, Cunard, the American Guion line - perforce followed his successful lead. Size, speed, and comfort all grew until the transatlantic passenger liner took on its final character of a high-speed, multi-storey luxury hotel. The Inman liners *City of New York* and *City of Paris*, launched in 1888 and 1889, were 10,000-ton steel ships with triple expansion engines, twin screws, five decks including a full-length promenade deck, and spacious, high-ceilinged dining saloons comparable with the most expensive restaurants ashore.

All this was good news for tugs, especially in Liverpool. Passenger liners working the Far Eastern routes, leaving from London or Southampton, were smaller because harbours east of Suez tended to be shallower; but they still needed tugs.

The new cargo steamer, too, quickly grew bigger in search of economies of scale. In thirty years the typical cargo liner out of Liverpool nearly trebled in size: in the 1860s it was around 2000 tons, and in the 1890s, 5 - 6000 tons or better. And size was only part of the story. Because ocean-going cargo steamers were faster than their sailing counterparts, they could make more journeys in the same time. Comparing the ordinary cargo steamer of 1900 with the best and fastest of the sailing ships of 1870, a steamer in a year could carry four times the amount of cargo three times the distance, at less than twice the cost. Although there were 5000 fewer ships, because fewer were needed, there were more ship movements, which spelled more work for tugs rather than less.

Another factor that increased the number of ship movements was the spread of submarine telegraph cables worldwide. The transatlantic cable was in continuous working order by 1866, and by 1887, more than 100,000 miles of other cable had been laid as well. Most cargo ships, whether sail or steam, were 'tramps' - the word coined in the 1870s to distinguish a ship that followed a tradition as old as sea trading itself, from a cargo liner with its fixed schedules and routes. The tramp, as its name suggested, sought any work, wherever and whenever it could be found. It carried cargo to any port, and there either picked up whatever other cargo might need to be carried to another destination, or sailed on in ballast to seek cargo elsewhere. Tramps carried most of the world's low-value, high-volume goods, slowly and cheaply when compared with the cargo liner. Their operations ceased to be so speculative and casual with the coming of the cable. 'Forward business' was now possible: the tramp ship's owners knew where it was and when, and where and when another cargo could be found. That cargo was negotiated before the first one was delivered. By making the tramp bustle, the cable helped to make business for tugs. And it helped tugs more directly, too, by allowing them to be called across an ocean to salvage a valuable ship without lengthy delays.

Tug owners adopted the technical advances pioneered on the high seas;

some of them very quickly, others after a delay. The spread of the surface condenser to tugs was delayed until the 1860s. They were expensive, and were needed only when tugs began to make long sea journeys and salt in the boilers became a problem. Tugs that made short sea trips could be blown down in their home ports between times, and those confined to dock and river used fresh or (at worst) brackish estuary water in their boilers. Since boats in the last two categories were in the great majority, the jet condenser survived a long time in the tug. The widely divergent demands of different kinds of towing meant that the screw also came late to tugs. Seeking, salvage, and ocean tows called for tugs that combined maximum power with maximum economy when running at a steady rate over long distances. Tugs for towing lighters did not need a lot of power, but they did tow at sustained speeds. In these respects screw propulsion had the advantage over paddle wheels; and since long ocean tows and lighterage were growth areas, from the end of the 1870s new tugs were usually screw-driven.

Not that screw design was straightforward. Any screw had to be submerged as fully as possible, to avoid waste of power, and hence fuel. A large-diameter screw offered better traction - pulling power - but needed a deep-draught tug to be completely immersed. A small screw gave more thrust, and hence more speed. Until the coming of the variable pitch propeller, which gave the best of both worlds, a compromise size was usual. Twin screws were the first solution to the problem; and they brought benefits of their own. They gave more traction, and more speed, than a single small propeller, but, being small, were able to work in shallower water - as when pulling-off a grounded ship - than a single large propeller. And since each was driven by a separate engine, twin screws offered greater manoeuvrability than any single screw.

Many of the factors that worked in favour of screw propulsion in tugs urged other innovations too. Any invention that promoted economy was good news, particularly if (at a time when tows were becoming more unwieldy by the year) it also increased structural strength, and either boosted power or permitted the higher boiler pressures that did so. Such were compounding, triple expansion, superheating, and steel construction of hulls and boilers. They were introduced gradually - for one thing, more complex machinery called for the expensive, time-consuming re-education of tugmasters and engineers with little or no formal education and a lot of inherent conservatism. But enough were in use by the end of the 1880s to show that it was possible to reconcile those hitherto difficult bedfellows, economy and power.

Some new technology was reflected in the iron paddler Watkins added to their fleet in 1862 - the *Annette,* named after William's wife - in that her engines had steam superheaters. So did the *Anglia,* another otherwise conventional tug. The *Era* of 1869 was more advanced. Watkins's first screw tug, she was a tiny, 30-ton boat for river work, drawing only 5 ft of water. With her compound engine she was notably economical, returning a consumption of 1¾ lb of coal per horsepower per hour. At the same time the *Era* was fast for her date, attaining 11 knots running free, thanks to a boiler pressure of 60 pounds per square inch and a power output of 104 hp. Her funnel could be lowered, allowing her to pass under London's bridges when towing lighters upstream from the Pool.

A new ocean-going paddler, the 209-ton *Cambria* of 1870, had big cabins for long cruises, and her after hold was convertible into a ladies' saloon, enabling her to be used as a tender. The *India* of 1876 was in the same class as the *Cambria*, designed to go seeking further down Channel than any other Thames tugs. In the same year the old *Monarch* was finally sold for scrap: Annette Watkins used the £40 she fetched to buy a silver tea and coffee service to remember her by.

A significant addition to the Watkins fleet was made in 1880: the *Canada*, a forerunner of the modern screw tug, in that she showed that power, economy, seaworthiness, and modest size could be reconciled in a manner unthinkable twenty years earlier. An 80 ft vessel of only 72 tons, yet a good sea boat, she was at first given a simple engine, but a compound was substituted in 1884; and she was converted again to triple expansion four years later. The larger, steel-built *Columbia*, of 1884 was another sea-going boat, with big bunkers and compound (later triple expansion) engines for maximum range. Watkins favoured the single screw; but they built one twin-screw steel tug, the 337-ton, 140 ft *Oceana* of 1888, with 1000 hp compound engines. At the time she was the biggest and most powerful tug on the Thames; yet she burned only 1 lb of coal per horsepower per hour.

Booming business on the Thames, and elsewhere, inevitably attracted more competition. The Black Ball Tug Company was run by a syndicate of Gravesend pilots; the tugs of the 'Ben' fleet were all named after Scottish mountains, being owned by Aberdeen shipowners. They were subsequently bought out by Watkins. More significant for our story were four other newcomers, all of whom, like Watkins, in due time became strands in the weave of The Alexandra Towing Company.

The Elliott Steam Tug Company was founded by a Thames pilot, Thomas William Elliott, in the early 1860s. In the old-fashioned partnership style of company, ownership of tugs, as of other ships, was commonly divided into sixty-fourths. In 1881, Matthew Dick and John Page joined Elliott in such a partnership. Dick and Page were shipowners and shipping agents; as agents, they managed the firm's tugs at Gravesend. From there, Elliott tugs competed in the seeking business.

In 1879, Charles Edward Gaselee, a London merchant, acquired the little 20-ton tug *Wasp* for towing the vessels of Thames lightermen. With the opening of London's new docks boosting their business, Gaselee had chosen

Entrance to London Dock.
(National Maritime Museum — Greenwich 58/5257).

'Gamecock', the vessel which gave its name to the Gamecock Steam Towing Company, run by a syndicate of Gravesend pilots (National Maritime Museum — Greenwich).

his customers well. In the first nine years of his business, Gaselee's jaunty, red-banded yellow funnels became a familiar sight on the river. He built up a fleet of five screw tugs, towing lighters between the Royal Albert Dock and the upriver warehouses. All but one - the *Dragon* - were named after stinging creatures: *Wasp, Gnat, Hornet, Scorpion*. The *Dragon* and *Scorpion* were 40-ton boats. The original *Wasp* was sold in 1888, and a new, bigger boat of 38 tons and the same name introduced.

The Gamecock Steam Towing Company, another firm run by a syndicate of Gravesend pilots, began operations in 1880. It used screw tugs from the first, the original one being the boat that gave the company its name - the *Gamecock*. The fourth newcomer was W.H.J. Alexander, a lighterman, wharfinger (wharf owner) and warehouseman who, like Gaselee, put his money on lighter towage from the new docks when he first acquired a tug. This was in 1883, when he built the 50-ton *Little England*, a 72 ft screw tug with compound engine and water-tube boiler. Alexander was a man dedicated to his work, who slept in a camp bed in his office overlooking the Thames at Wapping, so that he should not miss any early morning jobs that were going. Not that any owner of lighter tugs made a living by sitting in his office. Most work came his way in smoke-filled dockside pubs, where owners of tugs and lighters made deals over a pint, and foremen and men fed them with gossip and information.

As might be expected, trade for tugs expanded in booming Liverpool, too.

By 1865 the Liverpool Steam Tug Company had twelve tugs, and the Jolliffes were running eleven. From 1879 the Jolliffe tugs took on the family name: in that year the huge, 467-ton, ocean-going screw tug *Thomas Jolliffe* was launched. When built in 1885, the 975 hp, 149 ft *William Jolliffe*, although smaller at 332 tons, was one of Liverpool's most powerful sea-going tugs. The opposition was reinforced by, among others, the Liver Steam Tug Company in 1861, the Caledonian Steam Tug Company three years later, the Star Steam Tug Company, and - in 1877 - the arrival of William Becket Hill on the scene. He was freight manager of the old-established Allan line to Canada, which had gone over from sail to steam in 1856; and he would be made a partner in the firm. Nearly a century later, Hill's new enterprise became the second big Liverpool tug business to be absorbed by the Alexandra Towing Company.

The name of Hill's business - the Liverpool Screw Towing and Lighterage Company - signalled his enthusiasm for screw propulsion. All the 'Cock' tugs, as they were known from their names, were screw-propelled. And Hill was an innovator, too, pioneering twin screws and triple expansion engines in tugs. His first boat was the little 36-ton *Bantam Cock*, 68 ft long and developing 300 hp, built shallow-draught for river work. The 325-ton, 155 ft, ocean-going *Stormcock* of 1878, built by Laird, was an altogether different proposition. She was the first tug to be given twin screws, and her compound engines delivered 1000 hp. She could attain the then remarkable speed, for a tug, of 12 knots running free.

The *Stormcock* acted as a tender to the Allan liners managed by her owner. The combination of pulling power, economy, and reliability offered by her particular configuration commended her to the government, who chartered her as an auxiliary for the military operations in Egypt in 1882, when she was crewed by naval ratings and towed troops on the Suez Canal. She was then bought by the Admiralty. Two more tugs of the same name followed in succession in 1883 and 1885, as each in turn was sold into Admiralty service - a tribute to their qualities.

The *Gamecock* of 1880, not to be confused with the Thames boat of the same name, was a 155 ft steel tug of 401 tons, massively powerful with her 1250 hp engines, and also built for tender work. The 245 ton, steel-built *Blackcock* of 1885 followed the same pattern, with a single screw, but with the novel feature of triple expansion engines. These produced 1200 hp and propelled the *Blackcock* at 13 knots. For all their size and power, these early Cock tugs were elegant craft, with their twin raked funnels and masts, distinctive livery of cream funnels, and a cock weathervane at the masthead. More important, they were the finest representatives of their type, the most advanced and efficient tugs operating from British ports. When they were new and amazing, their superiority was acknowledged even by tugmen of other ports.

Their Prendiville contemporaries were just as good-looking, and almost as sophisticated. The *Knight of the Cross* and the 275-ton *Knight of St John* and *Knight of St Patrick* of 1883-5 were all powerful, twin-screw seeking boats with compound engines. Power went with size in the tugs of the Liverpool Steam Tug Company, too. Their screw tug *Blazer* of 1888 - the second of the firm's fleet to bear the name - was a 283-ton boat developing 1000 hp. *Pathfinder* and *Wrestler* were 1200 hp paddle tugs, bought from other operators. Boiler pressures were unusually high in some Prendiville boats. Another *Knight of the Cross*, bought in 1890 (the first had been sold to the Admiralty), had triple expansion engines and a boiler pressure of 160 psi. The *Knight Templar* of three years later was a sea-going tug in the modern idiom, modest in size but with boilers running at 200 psi. These gave 700 nhp - the same as the old *Anglia*, in a boat a third shorter, of not much more than half the tonnage, and incomparably more economical.

The owners of the big liners now proliferating in Liverpool, such as Cunard, needed more tenders. The first twin-screw 'Cunarder' was in fact the 612-ton tender *Skirmisher*, built of steel in 1884. She had a distinguished career, attending seven transatlantic speed record-holders in the Mersey, including (in later years) the *Lusitania* and *Mauretania*; and escorting the Royal Yacht *Victoria and Albert* at the opening of the Manchester Ship Canal in 1894 and the liner *Campania* at the Diamond Jubilee Spithead naval review three years later.

The kind of work done by tugs did not change between the 1860s and 1880s, though its scale did, and tug types continued to diversify in pursuit of it. Most owners preferred paddle tugs for big-ship handling in port and dock. The small screw tug was the coming thing in lighter work, which was booming, particularly on the Thames. In open-sea towing the bigger paddle tugs were busy, but by the end of the period they were being outclassed by the powerful yet economical screw tug with compound or triple expansion engines. The sea-going screw tug fell finally into its mould when the Gamecock Steam Towing Company showed the way to remove its last drawback. To prevent the excessive pitching that lifted a screw clear of the water, the company's new *Woodcock* of 1884 was given the full lines forward and steep upward curve of foredeck that endowed the tug with the bluff-bowed 'important' look that became its hallmark. The screw tug was an almost perfect instrument for this, the great age of seeking and salvage, of the first heroic, trans-ocean tows - before the sea-going merchant sailing ship went into decline; when steamers were multiplying (but before twin screws and steel construction had spread to enhance their reliability); before foreign ports acquired the floating docks and dry docks that made a tow home unnecessary; and after the laying of intercontinental submarine telegraph cables enabled tugs to be summoned quickly to far places. This was a new sort of salvage, done by pre-arrangement with insurers for a fixed

'Coburg', built in 1934. A detail of a painting by Gordon Ellis.

'Avenger', William Watkin's tug, built by Cochrane & Sons Ltd, Selby, in 1962. Acquired by The Alexandra Towing Company in 1975.

Sold in 1973, 'Canning' was the company's last steam tug. She went to the Swansea Maritime Museum in 1974 where she is still on display.

'Nelson' under power, pushing up white water. (J. Godsell, Wallasey).

A quiet moment for 'Sun XXVII'. She was built for W.H.J. Alexander in 1968 and acquired by The Alexandra Towing Company in 1975.

'Wellington' with heavy lift barge 'R.B. Brunel' in tow in the mid 1970s.

'Sun Kent' and 'Sun Essex'. Sister tugs built in 1977 for the fleet on the River Thames. (Fotoflite, Ashford).

'Alfred', 'Crosby', 'Albert' and 'Margam' tow out the 150,000 tonnes base of the world's largest offshore production platform at that time — from Loch Kishorn to Loch Carron, September, 1976. It was used by Chevron in the Ninian Field in the North Sea.

The Thames Barrier under construction in 1979 with a little help from Alexandra Towing. (Handford Photography, Croydon).

'Hendon' bought from France Fenwick in 1978. She was virtually new, never having been used by them. (Fotoflite, Ashford).

'Trafalgar' and her sister ship 'Nelson' were built by Richard Dunston in 1966, 1430 hp and 21 tons bollard pull.

The familiar skyline of Liverpool as a backdrop to 'Wallasey', sister tug to 'Sun London'. (K.G. Medley, Wallasey).

'Sovereign Explorer' proceeding down river with 'Implacable', 'Canada', and 'Collingwood' after her launch from Cammell Laird's Shipyard in 1984. This job takes a lot of muscle and skill.

'Sun London' towing 'Maidstone' class support ship for nuclear powered submarines. (Fotoflite, Ashford).

A reminder of the days of sail as 'Gloria' is handled by Alexandra tugs during the visit of the Tall Ships to Liverpool, 1984. (Hal Mullin, Bromborough).

Following their reception in the Pool of London, 1982 'Sun Thames' and 'Ganges' use their monitors for this special occasion and flags fly from Tower Bridge.

Greeted by thousands ashore and afloat, 'H.M.S. Warrior' 1860 entered Portsmouth Harbour 16th June 1987. The welcome included fireworks, water cannon and hundreds of balloons. 'Formidable' and 'Sun Anglia' towed Britain's first ironclad from Hartlepool to her permanent berth in Portsmouth Naval Dockyard.

'Formidable', 'Sun London' and 'Sun Anglia' assisting 'Mobil Petrel' to Coryton No. 4 Jetty in the Thames Estuary.

'Waterloo' another famous name of Alexandra Towing. Pictured here is No. 2 with that name.

'Atlantic Concert' with 'Bramley Moore', April 1986. (P.B. Parker, Formby).

In 1982 Alexandra tugs helped in the raising of Henry VIII's warship 'Mary Rose' within sight of Southsea Castle, Portsmouth.

Alexandra tugs manoeuvring two ACL container ships in the Royal Seaforth Container Terminal. Both vessels are just one metre shorter than QE2. (P.B. Parker, Formby).

'Waterloo' ahead, assisting 'H.M.S. Campbeltown' at her launch from Cammell Laird's Shipyard, River Mersey, in October 1987. (P.B. Parker, Formby).

'Margam', 'Alexandra', 'Herculaneum' and 'Fabian's Bay' assisting 'Cassiopeia' to berth at the Port Talbot Ore Terminal.

The Customs House and Canning Half-Tide Dock, 1880 with the 'Great Conquest'.

'Blackcock' the famous tug owned by Liverpool Screw Towing & Lighterage Co. Powered by a triple expansion engine producing 1200 i.h.p. seen here towing an Admiralty floating dock from Birkenhead to Portsmouth in 1912.

fee, not by chance encounter, a seized opportunity, or victory in a race.

In 1874 the Watkins tug *Victoria* was called from London to Lisbon to tow a steamer back to the Thames; with the *Anglia* she then immediately left again for Falmouth, to tow another steamer to Bremerhaven in north Germany, the *Napoleon* replacing her off the South Foreland. This kind of marathon set a new standard for tugs, and opened the eyes of shipowners to their possibilities. But already it was being eclipsed by more dramatic salvage work. At the end of 1870 the 2200-ton Lamport & Holt steamer *Biela* lay helpless at St Vincent in the West Indies with a broken propeller shaft. The Liverpool Steam Tug Company's *Cruiser* brought her home to the Mersey in thirty-two days. Watkins capped this epic in 1875, when the *Anglia* was sent out to St Helena to bring the broken-down, 2000-ton Union liner *Syria* into Southampton; a task which she accomplished, for the sum of £4800, in spite of her small bunkers and ravenous fuel consumption. It was the longest salvage operation so far completed by any tug, and only made possible by frequent refuelling. On the return 5000-mile journey with her tow, the *Anglia* had to coal at Freetown in Sierra Leone, Goree in Senegal, Tenerife, Madeira, and Vigo in Spain. The *Syria* had lain idle eleven weeks at St Helena, so her bottom was thoroughly foul, making her a particularly awkward tow. The voyage home took fifty-one days.

In the same year the Watkins paddler *Hibernia,* bunkers full and coal stacked on deck, was sent out to Madeira to bring home the broken-down steamer *Ariadne*. A storm in the Bay of Biscay stove in her hatches and flooded the engine room and stokehold. Her skipper nursed her, crippled, into Falmouth, and the *Anglia* was despatched instead. She completed the job, towing the *Ariadne* into Antwerp to discharge her cargo, and then to London for repairs, for a total of £1500. The next year, 1876, was a record one for the *Anglia*, earning her £12,700. Part of it was earned when she sailed east to tow a 3400-ton steamer from Port Said to Glasgow, for £2650. *Anglia*'s annual average earnings were something over £5000.

Not all tugs were so lucky. In November 1877 the barque *Charles Davenport* went ashore on a shoal off the South Foreland in heavy weather. The *India* pulled her off, but she then collided with another ship, ran aground again, and this time broke up. The tug lost the salvage, getting only £40 for her original service. A week after the steamer *Persian Monarch* ran ashore on Portland breakwater, her owners went bankrupt, and Watkins - who had salved her - had to 'arrest' her (legally seize her) in order to get their salvage award. Sometimes a tug saw no return at all at the end of the day. The steamer *Chateau Margaux* was reported overdue somewhere in the Bay of Biscay. The *Anglia* spent ten days searching for her in vain: a Cunarder had towed her into Irish port of Queenstown. This was in 1887, before the days of wireless, when there was no way of contacting a tug at sea. Tons of coal went to waste, and a tug might have been profitably employed elsewhere.

By 1885 more powerful, economical tugs were accomplishing long tows faster, and with less effort. This was particularly true of the sophisticated tugs of the Liverpool Screw Towing and Lighterage Company. The twin screw *Stormcock,* third tug of that name, towed the sailing ship *Ardencaple* from Fernando de Noronha off the coast of Brazil to Greenock, a distance of 4000 miles, in thirty days including only two coaling stops. This was no fluke. When in 1888, the big 4200-ton steamer *Norham Castle* lay disabled at St Helena, the *Blackcock,* steel built with triple expansion engines, towed her the 4500 miles to London in thirty-nine days, coaling only once, at Goree. A year later the twin screw steel tug *Gamecock* brought the *Yarrowdale* the 2560 miles from St Vincent to Dunkirk in fourteen days - less than half the time taken by the *Cruiser* for a roughly similar job nineteen years earlier.

Salvage could be lethally dangerous. In 1880 the New Zealand Shipping Company's sailing ship *Pareora* lay anchored in the Downs when hurricane-force winds parted her cables. She just escaped running on to the Goodwin Sands, and was being driven north on to the Gabbard Shoals, damaged and helpless, when the *Cambria* managed to get a line on board. The £2000 that the Admiralty Court awarded the tug reflected the difficulty and perils of the job. A salvage did not always end so luckily. Another New Zealand sailing ship, the *Famenoth,* was being towed out of the Thames to the Nore in March 1882 by the 'Ben' tug *Benachie* when hurricane force winds blew up. The *Famenoth* anchored, but even though the tug was holding her as well, she was swept on to the Pan Sands. Her passengers and crew were transferred to the tug; but when the ship's crew tried to return to her, four of them, and the pilot who went with them, were lost. When the weather moderated the *Benachie* returned to the *Famenoth,* and with the help of the Watkins paddlers *Titan* and *Punch* pulled her off the sands.

A year earlier, tragedy had struck at the Thames tugmen. The old Watkins boat *Napoleon,* which had been valuable enough to be modernized in 1878 at a cost of £4200 (more than many tugs cost to build new), went with the *Victor* to help the sailing ship *Allanshaw,* disabled in a gale off the North Foreland. The *Allanshaw* had lost her anchors and chains and most of her sails, and her windlass was broken. The *Napoleon* failed to get a rope aboard; but the *Victor* managed it. She held on, waves breaking over her, her master John McCarthy and his helmsman lashed to their bridge rails, from three o'clock in the afternoon until eleven-thirty that night. The rope parted; but at three o'clock the following morning the link was re-established. Finally, at eleven a.m. the *Hibernia* came up, and the *Allanshaw* and all aboard her were saved. The *Napoleon,* which had stood by the *Allanshaw* for eighteen hours in the teeth of the gale, was last seen from the *Victor* holding her own against giant seas. When visibility improved, the sea was empty. William Hough, the tugmaster, and his nine crew had vanished with their boat. The surviving tugs were awarded £3000 for the rescue, and John McCarthy was

The Cunarder 'Campania' with two twin paddle tugs standing by at Princes Stage at the beginning of the century.

commended by the Admiralty Court for his courage; but all that could be done for the men of the *Napoleon* was to raise a public subscription for their widows and orphans.

Other jobs were unusual because of the nature of the tow. The American millionaire Ross Winans was convinced that the future of the fast transatlantic passenger ship lay with his design, a cigar-shaped steel tube with a propeller at each end, and two tall funnels, a promenade deck, and a saloon 'somewhat resembling a railway tunnel' in between. The first of these enormously expensive aberrations was launched at Le Havre in 1866 and towed by Watkins's *Annette* to the Thames for fitting-out. Her sea trials were a disaster. Winans's creation was very slow, and pitched alarmingly in seas which hardly bothered the tugs. She was towed back to Le Havre in 1870 and sold for scrap.

A tow that was, if possible, even more unlikely came Watkins's way eight years later. 'Cleopatra's Needle' was in fact an obelisk erected in Egypt by the Pharaoh Thutmosis III nearly 1500 years before Cleopatra's time. In 1820 the ruler of Egypt, Mehemet Ali, presented it to the British government as a token of esteem, though it remained in Egypt. A similar obelisk was set up in Paris fifteen years afterwards as a monument to Napoleon's conquest of Egypt, and Cleopatra's Needle began its journey to Britain in September 1877, to be set up to commemorate Britain's subsequent reversal of Napoleon's Egyptian fortunes. The two hundred tons of rose-pink granite were inserted into a hollow iron pontoon in the shape of a cylinder, 92 ft long and 15 ft in diameter, on top of which was erected a deck house with a steering wheel and accommodation for a crew of three.

The steamer *Olga* towed the pontoon (christened *Cleopatra*) from Alexandria as far as the Bay of Biscay without incident, but ninety miles north of the Spanish port of Ferrol a storm tore her loose. Her crew were rescued by the *Olga*, but six men from the steamer were drowned trying, and failing, to secure the pontoon. It was eventually found drifting by the steamer *Fitzmaurice* and towed into Ferrol. The *Cleopatra* was not a big tow, but an awkward one, for which a tug was best suited. She tended to pitch badly, and to turn broadside-on under tow. For the sum of £500, the *Anglia* was commissioned to do the job that brought her into the public eye. In January 1878, after an uneventful six-day journey under the eye of an *Illustrated London News* artist, Captain David Glue towed the *Cleopatra* up the Thames between banks lined with cheering crowds. The schoolchildren of Gravesend were given a half-holiday to watch. The little *Era*, with her funnel folded down to negotiate London Bridge, took *Anglia*'s place for the tow from the docks to the Thames Embankment. There, in September 1878, Cleopatra's Needle was raised in its present position.

Glue was famous for taking his retriever dog with him on the *Anglia*. Off Newhaven, the dog was lost over the side. It was picked up off Calais, still swimming strongly, and returned to its owner with a silver collar recording the circumstances of the 'salvage'.

Less sensational tows than the *Cleopatra* gained less publicity, but were more difficult. In 1883 the owners of an 80 ft tall grain elevator wanted it floated across the Channel from the Thames to Antwerp. No tug operator but the resourceful Watkins would take on a job that was so far outside anyone's previous experience; but the *Hibernia* and *Titan* completed it successfully.

But, as ever, it was a useful volume of steady, unspectacular, routine work that a tug owner needed. Watkins were luckier than some in this respect. In 1877 they obtained the New Zealand Shipping Company's contract for towing their big sailing ships from London down Channel to the vicinity of Plymouth, for £95 a time. This was not a high figure, but it was regular, and got a tug into a good position for seeking inward-bound ships, so that it need not come home empty-handed. And in 1881 the P&O company returned from Southampton to London and, as promised, gave Watkins back their business.

There were still unproductive spells during which tugs and crews lay idle; and it was to help fill these that the operators of bigger, more seaworthy and comfortable boats began to seek charters for general passenger work, particularly from the wealthy and great. In 1873 the *Anglia* carried the visiting Shah of Persia and his suite during the Spithead naval review. The *John Bull* and later the *Cambria*, with her specially spacious accommodation, were chartered by the Duke of Bedford as yacht regatta tenders. The *Cambria* earned her owners £20 a day, temporarily exchanging her workaday black funnel and red-and-green banded hull for the yellow funnel and white hull of a yacht. Even her deckhand was kitted out in a smart uniform. He told the ladies of the ducal party that he was the duke's nephew, which was convincing until he was discovered in overalls painting the tug's funnel. Other tugmen were still more enterprising. Jack Spicer, master of his father's Thames tug *Undaunted*, heard that the Boulogne tug had broken down. Boulogne knew him well, for he ran summer excursions there, with passengers packed into his tiny aft saloon. Spicer nipped across the Channel to tow the local fishing boats out to sea, thereby incurring the wrath of the French authorities, who threatened him with prosecution for violating the law that reserved all such work for French vessels. Spicer used to ingratiate himself with the French by standing on a paddlebox and playing the 'Marseillaise' on his violin as his tug sailed into Boulogne.

Temporary jobs could be heartbreaking, as when tugs were chartered by the Customs and Excise to dump seized contraband tobacco at sea, under the watchful eye of Customs officers. Or they could taste of romance, without the danger that usually accompanied it. At the outbreak of the Franco-Prussian War the French government chartered the old Watkins paddler

Albion and the *Cambria* for £5000, to warn French ships in the Channel approaches. Each tug sailed with a French officer on her bridge, who was in the habit of wearing full uniform and decorations. Since Britain was a neutral power, this caused alarm at the Foreign Office. Watkins's tugs were also chartered by the Prussians, and the *Cambria*'s activities on behalf of the French did not save her from being detained at Cherbourg on suspicion of spying for their enemies. Cherbourg was not, in fact, healthy for the *Cambria*. She was seized there again in 1887, after towing a ship in from Dunkirk, for breaking the law of which Spicer had fallen foul. No owner or captain could have been ignorant of such a law. It was, no doubt, less of a deterrent than a challenge, especially if business happened to be slow elsewhere.

One of the more improbable jobs that came a tug's way was whaling. In 1865 the *Annette* sailed for Iceland waters on charter. A powerful tug could be useful in more than one way to whalers, whose ships were then still under sail, and whose whaleboats, that did the killing, were oar-propelled. Just before New Year 1884 the paddler *Iron King* was on charter from F. Warren of Newcastle, towing pontoons in the River Tay during the construction of the railway bridge replacing the one that had collapsed so disastrously five years before. News came that a 60 ft humpback whale had been harpooned in the mouth of the river, but that the Dundee whaleboats concerned were having trouble capturing it. The *Iron King* came up to assist, taking a harpoon line from a whaleboat; but the whale made off at high speed, towing after it not only the tug but two whaleboats and a steam launch as well. The beast was harpooned several times more; but night fell and it showed no signs of weakening. Breaking two harpoon lines, it crossed the bar of the Tay, making for the open sea, still with the tug and whaleboats in tow. It struck south until within six or seven miles of the Firth of Forth; then turned back northwards. When daylight came the whalers' harpoon gun fired another harpoon, two marlin spikes, and an assortment of nuts and bolts into the quarry until the sea was red with its blood; then the ammunition ran out. All to no avail. At 8.30 in the morning the last harpoon line parted, and the whale disappeared, still with three harpoons and various other ironmongery in its body, leaving the *Iron King* puffing lamely in pursuit.

The *Iron King*'s exploit may have been legitimate, in that the tug's charterers had had no work for her at that particular moment: history does not relate. Or, it could have been a case of a tug out on a precious, long-term charter, lost to the customer for a good two days while the crew, perhaps too drunk to care, burned quantities of coal to no purpose. Masters got the sack for much less.

Life for the nineteenth-century tugman was no jolly adventure. When a boat was in dry dock for repair, the crew was laid off. Spells of idleness - which might end in lay-off if no work turned up - alternated with heavy, often dangerous and prolonged labour that lasted until the job was done, however long that took. Generally, only a crew kept on the docking and undocking of ships worked regular hours, and even then, the hours worked were likely to be what is now called 'unsocial', since they depended on the tides. If it was a long job, the crew of one of the smaller port tugs, numbering only six or seven men, worked on without the chance of a proper sleep - four hours on, four hours off. And if the work demanded it, all routine went by the board. While seeking, the master would be on his bridge almost all the time, completely exposed, battered by everything wind and sea could throw at him. If he left his post, his tug might suffer the fate of the Watkins boat *Bristol*, run down and sunk by a steamer when her master left a deckhand at the helm while he had breakfast below. Conditions below were not likely to be much of an improvement. The stokehold was known simply as 'Hades'; but in an old, patched tug it was hell with a difference, for the sea was in there as well as the fire. A fireman was likely to have to put on his seaboots to pump out the stokehold before he could raise steam.

Bigger tugs carried a crew of eight or nine: master, mate, second mate, engineer, leading engine room hand (not found in small tugs), two firemen instead of one, deckhand, and boy. With a large crew, it was possible - conditions permitting - to work three-hour watches with six hours off. Cable laying in the narrow seas was a popular job, because bad weather stopped work, and the crew could spend nights in port: a rare luxury when at sea. On long hauls, there would be extra hands on board. There might be one more on deck and one below, together with a second boy as trimmer to keep the bunkers evenly stowed. Tugs on long-distance salvage jobs might well work outside home waters, defined as beyond the Elbe mouth in the North Sea and Ushant at the western tip of Brittany. A tug bound for more distant waters had by law to sign on a certificated master: tugmasters who themselves had certificates were uncommon, for tugs did not usually work outside home waters. From 1851 all newly appointed masters were obliged to have a certificate of competence from the Board of Trade, won by examination, but the so-called 'sailing master' appointed to a tug was usually a retired ship's captain with no tug experience, whose navigating skills were rusty, or suspect. No master who held a certificate awarded before the regulations came into force was obliged to take the examination, so improvement was slow. Two-thirds of ship losses between 1856 and 1872 were held to be due to the drunkenness, ignorance, or incompetence of masters. One had been known to mistake the lights of France for those of England. The sailing master was resented by the tugmaster; the more so since he got 10s a day plus his keep for the duration of the voyage - double what the tug captain was paid. Sometimes the sailing master was content to let the tugmaster do his work for him; although this was illegal, and if he was found out, the tugmaster would be fined.

In the 1870s the master of a seagoing tug might be paid 35s a week, or

30s on a river tug. The mate got around 30s, the second mate 24s, a deckhand or ordinary seaman 20s, and the boy 12s 6d. There were no overtime payments, in spite of sometimes enormously long hours. Out of their wages, the crew had to find their own food: the tugowner paid for food only when his boats left home waters. The crew usually put 1s a day per head into a pool, or up to 5s a week in the case of the boy; and each took it in turn to do the catering ashore. One man got as far as buying a barrel of salt, before succumbing to temptation and drinking away the rest of the food money. By the 1890s the master of a seagoing tug could be earning £2 a week; and ten years later, might be doing so even on a little lighter tug. But his mate and engineer still got only 30s, and the deckhand 20s. The boy, however, might now be paid 15s a week.

As well as their wages, most masters and some engineers received commission, at the rate of 2½ per cent on their tug's takings, to encourage them to find business for their boat. Although an officer on commission might discover that his wages had been cut to compensate, there is plenty of evidence that commission made all the difference to an officer's earnings: in normal times, he could expect it to increase his income by about 50 per cent. The seamen got no commission on top of wages; but in the 1890s at least one Thames-based operator paid his men no wages at all, only a share of their tug's earnings. They appeared quite content with this extra hazard in their lives, for in good times, they said, they did very well.

The master might also own shares in the tug, and would therefore receive a share of any distributed profits of the business. Beyond tips for taking a pilot ashore, which were pooled, the crew got nothing extra by way of regular income. They prayed for salvage work, to win a share of the award; or for some rare special reward, as for life-saving, or for towing home a victorious tea-clipper. The proportion of a salvage award that the tugowner gave his crew varied, and was a bargaining counter in wage disputes. Early in the twentieth century The Alexandra Towing Company's crews seem to have divided between 15 and 20 per cent between them; on the Thames rather later, the most usual figure was 22½ per cent.

There was not much except hope for the deckhand to live on. In Liverpool, he could look down on the lighterman on his 'flat', for even its captain got no more than 30s a week or so, and its mate 25s; but on the Thames by the 1890s, the ordinary lighterman or waterman made more than the Mersey flat captain. The deckhand of a tug in home waters also did less well than his equivalent on a merchant ship. An ordinary seaman under sail or steam might get only 12s 6d a week in the 1850s; but he also had 1s 6d or so a day for his keep, making a total of (say) 23s. By the turn of the century he was receiving a basic 14s or 15s a week under sail, and perhaps 20s on a steamer, plus his keep. Of the other men working in the tugman's environment, only the dock labourer was everywhere worse off. If he worked full time, a London or Liverpool docker could earn around 30s a week in the 1890s; but on the Thames only 6 per cent of the workforce was permanent - the rest were casual labour, who on average worked for about half the year. In Liverpool work was better paid, but still more irregular. These men and their families were driven back on charity to survive.

All on board a tug, from the master down, were likely to have worked their way up from boys, the 'lowest form of life'. The owner's family was not necessarily exempt. W.H.J. Alexander's six sons all became tugmasters in his firm, but all worked their way up from boy, or lighterage apprentice, starting probably at fourteen. The boys did all the dull and menial jobs - cooking, washing up, cleaning the crew's cabins and the brightwork, carrying coal for the cabin fires. At the same time, an ambitious boy got in as much deck experience as possible, learning tug handling skills from the master and mate when they could spare the time. Usually they made time, for they had come up the same way. The boy knew that every boy was a potential master, who might get his own command by his early or mid-twenties; quite possibly he had seen his father, elder brothers, and uncles climb the same ladder before him. William Simmons found himself second mate of the Watkins tug *Nubia* at the tender age of fifteen, little more than a year after following his waterman father on to the Thames. Thomas Physick joined The Alexandra Towing Company at fourteen, was a master at twenty-one, and held that rank for more than half a century. Some masters stayed with one tug. S.R. Reader of the Watkins tug *Victoria*, appointed in 1877, remained in command of the one boat for twenty-eight years.

If a man kept his life and limbs intact, he was likely to serve his owners long. J. Thompson, a Watkins engineer, started as a boy in the *Victoria* in 1857 and retired in 1929, after seventy-two years' service. It was commonplace to stay with the same company for forty years. On being pensioned off, one man with fifty-three years' service with the same Liverpool towing company grumbled to his employer that 'If I'd known that it wasn't going to be a permanent job, I would not have taken it on'. Family tradition was strong. Alexander's sons detested one another; they were constantly having rows, or were in hot water with their father, but they went on working for the family firm. At one time, the Physicks, Newalls, and Howarths accounted for more than a third of the Alexandra Towing Company's Liverpool staff.

The dangers of the parting two rope, and the tow that overran or girted its tug, were always in the back of the tugman's mind. There were more bizarre incidents as well. In 1854 the Southampton tug *Belmont* was overrun by the ship *Walter Hood*. As she was overwhelmed, losing her funnel and mast on the way, her crew managed to scramble on board the tow. The *Belmont* then righted herself, and steamed off down Southampton Water on her own, her crew in futile pursuit in a rowboat. The men of the Calshot

Unusual sight at Gravesend during the great freeze of 1895 when the River Thames froze over. (National Maritime Museum — Greenwich).

lightship eventually got on board the tug, allowing her own crew to catch up and retrieve her. The *Belmont* was rammed and sunk by a steam collier later the same year, again without loss of life; and as usually happened when a tug sank in shallow water, she was successfuly raised and put back into service. The *Belmont*'s adventure had its comic side. Not so the fate of the Elliott tug *Revenge,* which fouled an unexploded shell off Dungeness in 1900. It blew up, killing three of the crew.

Insecurity, ever-present danger, and long, irregular hours of hard labour tended to produce a rough, foul-mouthed breed who did not look beyond tomorrow, who had no respect for authority or the law even when sober, and who, when drunk, went berserk.

One of Alexander's sons was Charles, a master for twenty-seven years. Once when he was a mate, his tug was waiting with a string of lighters below London's Hammersmith Bridge for the tide to fall sufficiently to allow them beneath. Among the crowd gaping down from the bridge was a very green young police constable. To relieve the tedium, Alexander let off his steam whistle in their faces. As he started back with the others, the policeman lost his helmet, which fell on to the deck of the tug. Peremptorily, the bobby demanded his headgear back. Alexander told him to address him as 'Captain', and lectured him on the virtues of good manners. The tugman tried on the helmet, which came down over his face and beard. The crowd howled with glee. The mortified policeman finally produced the polite request demanded; upon which Alexander threw the helmet back. It landed in the open top of a passing horse bus, and the constable was last seen sprinting in pursuit, still shouting for his helmet.

When drunk, the nineteenth century tugman was a menace. The Glasgow tug *Flying Childers,* master Captain McDougall, lay in the harbour of Ramsey, Isle of Man, in July 1860, with her tow the sailing ship *William Kidson,* which she was towing from Liverpool to Troon on the Firth of Clyde. While the tug was taking in coal, her crew had been taking in large quantities of alcohol. When she got under way, all hell broke loose. Her paddles fouled and damaged one fishing boat and set another adrift, the crews exchanging insults the while. After much ponderous manoeuvring, McDougall tied up to a moored schooner and led most of his crew ashore across her deck. On it the tug crew started 'a fierce and bloody fight' among themselves. They were eventually separated by onlookers, who dragged them on to dry land.

Meanwhile the tug's firemen, who had been left aboard, cast off and got her under way, setting off up the harbour with one of their number on the bridge issuing orders. She shortly reappeared, steaming straight into a cluster of small boats and demolishing one of them. By this time a large crowd had gathered on the quay. McDougall muttered curses but did nothing; the harbourmaster decided that enough was enough. He sent boarders on to the *Flying Childers,* where, after a brief battle, the firemen were subdued and the tug was secured. At this, McDougall roused himself and stormed back on board his boat, followed by the rest of his crew. General mayhem ensued, with scores of townspeople jumping aboard to join in the fray. Spectators noted that certain conventions were observed, however: one tugman who produced a knife was disarmed and turned over to a policeman, as was another who was laying about him with an engineer's turnkey - a sort of enormous spanner.

At last the tugmen were immobilized and led away to jail in handcuffs, bleeding, eyes blacked, jaws and cheekbones cracked. Their boat was impounded at the Customs House Quay. The magistrates could not wait to get rid of them. After a few hours in the cells they were all fined a nominal shilling, plus a modest-sounding £7 for damages and costs which Captain McDougall was required to produce, and turned loose into McDougall's charge.

5

Alfred 1901

A New Contender

By the standards of nineteenth century Liverpool, trade was slack in the dozen or so years up to 1887, the tonnage passing through the port rising by only 18 per cent, to 7.6 million tons. Such uncharacteristic sluggishness was shaken off in 1888, and in the next decade growth accelerated again. There were hiccups on the way, including a recession that lingered from 1893 to 1896, but by 1897 the tonnage handled on the Mersey had gone up by another 25 per cent, to 9.5 million tons in the year; and this was only the beginning of a new surge.

More than twenty major shipping lines carried goods and people in and out of the port. Passenger liners went on growing in size faster even than merchantmen. The Cunarders *Campania* and *Lucania* of 1893, 13,000-tonners disposing of 26,000 hp, inaugurated a new, spectacular, and final phase of competition for the Liverpool passenger trade, which lasted until 1914. The standard they set for speed lasted longer than that for size: their 22 knots was still a normal cruising speed for crack transatlantic passenger liners in 1914, for the economic limit had been reached. As ever, large size, rapid turn-round, and competition between customers were all good news for tugs. To handle the traffic, Liverpool in 1893 had 363 acres of wet docks and 24 miles of quays, Birkenhead's wet dock area was now 161 acres, and by 1896 dredgers had finally cut a passage through the bar across the river mouth that had hampered traffic for centuries.

One threat, as the shipowners saw it, to the promise of ever-sunnier prospects was a strike of seamen in 1889. It was part of a wave of stoppages by sailors and dockers that hit Liverpool, London, Cardiff, and Southampton in the years 1889-91. This one was led by the National Amalgamated Sailors' and Firemen's Union, formed in 1887 by James Havelock Wilson to better the pay and conditions of ships' crews. The response was such that in two years, the union had thirty-six branches and 40,000 members - perhaps one in four of British merchant seamen. This union, Ben Tillett's Dock, Wharf,

Riverside and General Labourers' Union of 1889 and the National Union of Dock Labourers - born on the Clyde in 1889, with its first branch on Merseyside the same year - were in the forefront of the militant 'New Unionism', which took up the cause of the mass of unskilled, unorganized labour, and sought confrontation with, and the downfall of, capitalism. In these respects it differed from the traditional, skilled craft unionism represented by the Amalgamated Society of Engineers, whose philosophy was to work in co-operation with capital for mutual benefit.

The main concern of the new unions was to increase and safeguard their members' employment. The employers accused them of restrictive practices that created work, over-manning, and the spinning-out of hourly-paid jobs. They fought the use of non-union labour, of 'free' labour imported by employers during strikes, and of labour-saving technology such as the grain elevators that had been introduced in Liverpool, Glasgow, and London in the 1880s. By means of steam-driven bucket conveyors, these machines unloaded grain ships faster and more cheaply than gangs of dockers working by hand in the holds.

The 1889 Liverpool strike failed. It gained the seamen nothing, and lost many their jobs, which were taken by non-union labour brought in by the shipowners to work their idle vessels. There would not be another major confrontation in Liverpool for more than twenty years. The greatest weakness of the New Unionism was the ever-present pool of cheap labour clamouring for employment - the casual labour of the docks, and the foreign seamen, European and Lascar, who by 1910 made up nearly 30 per cent of the crews of British ships, and half the firemen. The sailors' union was very vulnerable to trade recession, during which men already in want would work in any conditions. When a recession arrived towards the end of 1892, and continued until a million tons of shipping was laid up, the union's membership fell to 18,000.

Another threat that turned out to be more apparent than real was the opening in 1894 of the Manchester Ship Canal and its associated docks at Manchester. The canal, 35½ miles long and 38 ft deep, linked the Mersey with the heart of the industrial North, affording ocean-going cargo ships direct access to the markets for imports and sources of exports that Liverpool had hitherto monopolized. Naturally the canal was bitterly opposed by the Liverpool dock interests, and by the railway companies. No longer would it be necessary, they thought, for goods to be unloaded on the Mersey for onward shipment by rail, or railed to the Mersey for loading there. A major handling operation, with its attendant rail freight charges, would be eliminated. The canal was a great success - traffic on it grew from 930,000 tons in 1894 to 5.8 million tons in 1913, by which time Manchester was Britain's fifth busiest seaport - but Liverpool flourished regardless. The continuing growth in the size of merchant steamers meant that the biggest could not use the canal. As a result, some goods that had gone by rail were now transhipped to canal barges. And the Canal linked more than Manchester to the sea: to Liverpool and Birkenhead now came fleets of barges with salt and chemicals from the Cheshire mines and chemical works. The quantity of goods transported by barge rose from 239,500 tons in 1894 to 323,000 tons in 1913.

All this was good for Mersey tugowners, who had the means of towing the barges; but the opening of the canal brought ship towing work as well. A ship negotiating it had to move so slowly, to avoid damage either to the canal bank or itself, that it could not be steered, and so needed a tug. The Manchester Ship Canal Company became a tug operator. It chartered tugs from Mersey owners, and in 1895 bought its first tug, the little 94-ton *Mercia,* from Watkins. In the same year it built a tug specifically for canal work, the *Charles Galloway*. In addition, Mersey tug operators themselves worked on the canal. The big Mersey ocean-going tugs *Stormcock* and *William Jolliffe* caught the public eye in 1890 with a competition for the name of most powerful tug on the river, including a well-publicized 'race'. But from the late 1880s there was a great and growing amount of lucrative work for the smaller, less spectacular dock and river boat, handling the new tonnage on the river and working on the narrow waters of the Ship Canal.

One man who looked well placed to exploit this market was George Bell Cowl. He was a lawyer by training and a notary public by profession - an unusual occupation for a tug owner - who had started tug operation in Liverpool in 1882. By 1887 he had a mixed bag of five boats, all bought second-hand. Apart from being screw-propelled, they had few technical characteristics in common. Two - the 84-ton *Flying Tempest* and 88-ton *Flying Whirlwind* - had been Clyde-built in 1875 with compound engines and the remarkable feature of a screw and rudder at the bows as well as the stern. This, it was thought, gave more pulling power than a conventional layout, as well as better manoeuvrability. Certainly the tugs gave good and long service. Cowl's oldest acquisition was the tiny 29-ton *Turbot,* which had started life as a sailing trawler in 1866, had been converted to steam seven years later, and adapted as a tug by Cowl in 1883. The *Flying Breeze* and *Flying Kestrel,* Clyde-built in 1874 and 1878 respectively, seem to have been more normal craft. The *Kestrel* was a single screw boat of 135 tons. The 88-ton *Breeze* sank in the docks in 1886 while towing a Johnston liner. She was subsequently raised, though at considerable expense.

Cowl's business was not doing well: in August 1887 the *Flying Kestrel* was mortgaged, a straw in the wind. In the same month Cowl's tugs were taken over by a new business with an altogether more solid and professional look to it - the Alexandra Towing Company. This was a public limited company, called after the Mersey dock of the same name. Everything about the men who were the motive force behind the company spoke of the new, expanding

Liverpool, bursting with enterprise, single-mindedly dedicated to growth. Their experience and know-how were tightly relevant to that growth, and had helped to create it. Between them they represented a network of related interests, all dock-based, for which diversification into tug work was a natural progression which, by accident or design, came just before the latest surge in the city's growth.

The prime movers in the business were mostly Methodists, members of a persuasion famously imbued with the Protestant work ethic, with businesses and homes in Bootle, the fastest-expanding part of Merseyside. Bootle was a recently incorporated town - lately a village - immediately to the north of Liverpool that was in the course of becoming its suburb, with a frontage on to the Mersey that was rapidly filling up with the northward spread of Liverpool's deep-water docks. The new company moved into Cowl's old office at the Langton dock, Bootle: right at the scene of the action.

The Alexandra Towing Company was launched with a capital of £25,000 raised in £5 ordinary shares. The issue was over-subscribed. The chairman was William Poulsom, lately Mayor of Bootle. A first-generation Liverpool man, by trade he was a stevedore, servicing the Warren, Guion, and other lines, and a master porter - one of the largest employers of such labour in Liverpool. (Stevedores were the dockers who loaded and unloaded cargoes; porters were the warehousemen who shifted, stored, and despatched them: the names were used of both employers and men). Poulsom ran a similar business in London. He had also introduced the grain elevator to Liverpool, Glasgow, and other ports, and was the founder of the Grain Elevator Company that owned and operated them. Imported grain was big business in Liverpool and Glasgow, where by 1897 two elevator companies with thirty-seven elevators were discharging nearly a fifth of Britain's imports, and moving it with a fleet of thirty-seven barges. For nearly twenty years Alexandra board meetings and company meetings were held in Poulsom's Bootle offices.

Frederic Winbolt Staveley, of Staveley & Company, shipping agents, was a director of the new concern, and also of the Grain Elevator Company. In view of his trade and connections, it seems likely that his main function was to find business for the firm's tugs. Charles Kerr, another director, was described as a book-keeper by profession, and probably performed the functions of treasurer. Captain Thomas Durrant, who was appointed Marine Superintendent - the officer of a tug company responsible for the day-to-day waterside working of the fleet - seems to have been the only member of the board to have commanded ships and seamen. He had come up through sail in the coasting trade and the Australian clippers, had served nearly twenty years with Cunard, skippering seven of their ships, and had been Marine Superintendent of the Guion line until it collapsed. He had also been a director of the Liverpool Steam Tug Company.

George Bell Cowl joined the new board, no doubt to contribute more experience of tug operation, which was lacking in all his fellow directors except Durrant, and to bring in business with his existing contracts. At an early stage he was allocated 400 shares, and was also paid a total of £1750 over the first few months of the firm's life. The only important officer of the company who was at first neither a director nor a shareholder was John Griffith Hughes, appointed Superintendent Engineer in October 1887. He was responsible for maintaining the existing tugs, and later for supervising the construction of new ones. His appointment, which filled another gap by providing expertise on technical matters, completed the senior management team.

Its status as a public company allowed the new business to solicit funds from the general public, and so avoid, it was hoped, one of the pitfalls of tug operators - inadequate capital. It also allowed its promoters to raise these large sums without recourse to their own pockets. It did, however, mean that because the size of individual shareholdings in relation to the whole was modest, financial control could not be formally vested in a few people. Personal shareholdings were small - even Poulsom had only 220 shares to start with, and Durrant 300.

But it was still possible for an informal group of shareholders, such as a family, to wield considerable influence if they acted together, even when no member of it was on the board. On this basis, the most significant shareholding - about 1400 - was by early 1894 held by members of the multitudinous Mack family: ten of them in all, out of a total of thirty-six shareholders in the company. James Jermyn Mack was senior partner of James J. Mack & Sons, shipowners, ship brokers, and forwarding agents of Bootle. A Liverpool man, he had, like Durrant, worked for the Guion line. James Mack was also founder of WS & IA Mack, another big firm of Mersey stevedores and master porters. His brother Isaac A. Mack and son William Durrant Mack were partners in both businesses, and all three were eventually shareholders in The Alexandra Towing Company. Isaac Mack was a director of the Adelphi Bank, who were the company's bankers for a time from 1892. In addition to their equity interest, William Mack and three other members of his family, as well as other shareholders, lent money to the company against debentures, thereby increasing their influence over and commitment to its fortunes.

One of the largest individual shareholders in the company was Alexander Bicket. He was not on the board either; but his ties with the chairman, and with the river, were close. He had worked for William Poulsom since he was seventeen, and now, at the age of thirty-three, was a partner in the firm. They had other things in common, both being first-generation Liverpool men; incomers who had succeeded. Poulsom was a Wiltshire man, and Bicket was an Ayrshire Scot of Presbyterian family. Bicket, together with

Durrant and Kerr, was voted £300 for 'services rendered in connection with the formation and incorporation of the company'.

From the outset, the firm sought all the different kinds of job on offer, subject only to boats being available. Its bread and butter was, and remained, dock and river work. The banded black, white, and buff funnels of the new company became a familiar sight on the Mersey, towing lighters, canal barges, and ships. They looked for work not only in the north docks - from the Princes Dock northwards, where Cowl had operated - but in September 1889 extended their operations to the south docks too. Contracts - usually for twelve months, at fixed rates - were made with Cunard, for towing the 'flats' that brought coal to their liners; and with White Star, Lamport & Holt, Leyland, Bibby, Anchor, Dominion, and other lines. Needless to say, Alexandra's tugs also towed the grain barges of the Grain Elevator Company. Towage rates varied. In the good times of mid-1890 most operators charged £4 to £6 per tug between berth and river, depending on the size of the ship.

Alexandra's earnings from dock and river tows, as opposed to other jobs, are impossible to assess in the first seven years of the company's life, as they are not usually differentiated in the records surviving from this period. But little other work was done, apart from the occasional salvage, so for practical purposes the figures we have are probably good for dock and river work. The first full six months for which figures of any sort are available show an average of £217 a week, or about £43 per tug.

Naturally, the company's tugs took all the salvage they could find, even though the gains were highly unpredictable. The first came their way within a month of the firm's founding, but the master of the *Carl Rosenius* denied that he had needed salvaging, and Alexandra had to resort to their solicitor to extract the £60 claimed. This was fairly usual. In another battle, Alexandra claimed £800 for towing the barque *Robert* off Great Burbo Bank in Liverpool Bay, but settled for £275. No doubt the company always claimed more than they expected to win, but on this occasion, at least, they appear to have come off worst. The crews seem to have done well, taking home about 20 per cent of the award between them, the equivalent of a week's pay each even after modest awards of £160 - £170. But for Alexandra, salvage did not come in often enough to make it truly lucrative for anyone involved. In the company's first nine years of working, they won salvage jobs at the rate of less than one a year, and might earn as little as £12.

Two other kinds of work came the company's way. By 1890 they were tendering passenger liners. Four years later they were charging liners of the Canadian Beaver line 30s a ship when their rivals were asking 40s. But for the present, Alexandra did not invade the tender market seriously. The same could not be said for the Manchester Ship Canal. Well before it opened, Alexandra obtained permission to tow on the canal. By April 1894 they were towing Lamport & Holt's steamers on the new waterway, and in 1895 were generally earning a modest £30 to £60 a week from this source.

The company's business was in river and docks, not the open sea. In these early years ocean towing was neither sought nor accepted, unless - as seldom occurred - tugs could be spared from the Mersey. In any case, the former Cowl tugs were almost all small, and lacked the punch of the modern, powerful boats of the likes of Jolliffe and Cock tugs. Being less suited to sea work, they tended not to be offered any.

A sea tow was an event when it happened, the first known being the towage of a barque from Liverpool to Barrow-in-Furness by the *Flying Kestrel* at the end of 1890, more than three years after the company's birth. In February 1894 there was so much business in the Mersey that all tugs at work 'outside' were ordered back. Later that year came the first sign of a change in policy, when 'sea work' began to feature separately in the weekly returns of earning, but on the rare occasion when 'sea' earnings were over £100, the Mersey takings were at their lowest, bearing out the board's view that sea work lost port work. There was more business in Liverpool's river and docks than the five ex-Cowl tugs, hard-worked and constantly in need of repair, could handle. It was only four months before Alexandra, with the money behind it to pay for capital expenditure, was adding to its capacity. A new tug, the *Alexandra,* was ordered. She was a 197-ton, 250 hp twin screw boat named, like most of her successors, after a Mersey dock. She cost £4500, paid in instalments for the sake of the cash flow of both builder and client.

Two second-hand tugs were bought in October 1889 - the little 50-ton *London*, which cost £1600, and the powerful twin screw, 208-ton *Conqueror*. Another new tug was ordered early in 1890 - the *Hornby*, a duplicate of the *Alexandra*. In mid-1892 the single screw *Langton* was ordered, modestly sized at 92 ft and 157 tons, but extremely powerful with her 600 hp. A second-hand paddle tug, the *Nelson*, was added to the strength, and the tiny old *Turbot* was sold for £100. The *Nelson* turned out to be a mistake, and was sold two years later for half what she had cost - an exceedingly rare instance of Alexandra losing money on a tug.

Trade had begun to fall off after mid-1892, but Alexandra continued to spend on new, more powerful tugs. Ships in the docks were getting bigger all the time, and the extension and improvement of the docks continued throughout the recession. Furthermore, an operator with a capable, modern fleet would win the lion's share of what custom was available, and when better times came would be ready to take advantage of them. Although trade was slack, Alexandra had the cash to invest. Their name was good. In March 1896 they raised another £10,000 of share capital with a new flotation, bringing the total figure up to £35,000. Their borrowings were repaid punctually and they had no difficulty in raising loans, from the Grain

Elevator Company among others. In all this, Alexandra stood in contrast with the generality of tug owners, who lacked resources and foresight, and led a hand to mouth existence by comparison.

Another of Alexandra's few mistakes, the *London*, went in 1893 in part exchange for a new boat, the *Canada*, delivered at the year's end. Like the first three new tugs, she was built by S. McKnight of Ayr. In 1894 the United Steam Tug Company, one of the weaker brethren, sold two of their tugs and thereby forfeited the custom of the Bibby and Anchor lines. Alexandra stepped in, buying one of the tugs, a paddler, for £800, renaming her *Brocklebank*, and securing the towage their rival had lost. As the *United States*, the United tug was an antique dating from 1855, and had helped in the salvage of the *Royal Charter*; but she was still valued in the docks. Finally, in January 1896 the company ordered the *Huskisson*, a duplicate of the *Langton* and *Canada*. When she was delivered - the first Alexandra tug to come from John Cran of Leith - the company found themselves with a fleet more than doubled in numbers over nine years, incomparably more efficient, and already (in 1895) worth nearly two and a half times as much.

Financially, the first five years of the Alexandra story were a time of unbroken success. The next four coincided with the shipping recession of 1893-6 and were less cheerful, but because Alexandra was a well run and well funded business its rise was merely slowed temporarily. In these ways its progress mirrored that of the port it served. The first half-year's report set the pace. There was a net profit of £610; all repairs to the fleet had been paid for out of revenue; a dividend of 2s 6d on paid-up shares and 1s on ordinary shares was declared; and the directors awarded themselves fees of £125 each. An interim dividend of the same amount followed in April 1888. There were Christmas presents for the tug crews. Such good tidings became routine at half-yearly and annual general meetings of the shareholders. In the slack years, the company was surprised at its own success. In the second half of 1892 receipts were almost up to 1891 levels, £10,579 as against £10,761. The first six months of 1893 were 'very trying', but better than expected: earnings were down, and so were profits. As for the second six months, although the shipping slump went on, revenue was greater than in any previous half-year.

The picture changed for the worse in the second half of 1894, giving the board 'an unusually anxious time'. Revenue was sharply down again, but there was relief that things were no worse. Net profits were even marginally up, so profitability had actually improved. The recession dragged on for another two years, during which time the company trod water. It spoke of being able to 'maintain its position' and of 'fairly satisfactory' receipts. By the end of 1896 the sun had broken through. The company's tenth annual general meeting was told of a 'gratifying improvement' in business - receipts were up to £14,130 - in spite of three serious accidents to tugs that could not, as was normally the case, be paid for out of revenue. The *Brocklebank* was so badly damaged by a steamer that she almost became a total loss; the *Alexandra* was rammed by a steam barge, which sank; and the *Flying Whirlwind*, towing a ship on the Manchester Ship Canal, damaged herself running aground and had to be beached to stop her sinking.

The figures quoted - like those for the wages of tug crews - should be put in the context of their time to give a realistic picture. In the early 1890s prices were lower than they had been at any time in Queen Victoria's reign. The purchasing power of the pound was around forty times what it was in 1987, with income tax in the 1890s ranging between 6d and 8d in the pound.

Alexandra's early success was won by keeping revenue up and costs down, which was never less than a struggle, especially when it came to keeping their tugs afloat and in one piece. Making good damage to the fleet was a constant drain; and a tug in dry dock was not earning. In another sort of accident someone else's vessel was damaged, usually the tow. Its claim was almost always fought, but in most cases some compensation had to be paid, for the tug's crew was generally at fault to some extent. The sum was not usually large - £30 here, as little as £4 10s there - but they were many, and mounted up. More than damage could be involved: if a tug's negligence delayed a ship, its owners might claim demurrage, or compensation, and there was goodwill at stake, as well as cash. July 1889 saw the Cunard Company moved to complain officially about the 'inattention and incompetence' of the masters of the *Flying Whirlwind* and *Alexandra*.

Tugs hit their tows, or more often caused them to hit a variety of unyielding objects - other vessels, piers, dock gates, walls. They hit buoys, they hit tugs of other fleets, and they even hit tugs of their own fleet. They hit, more frequently than anything else, the flotsam and submerged rubbish that abounded in every busy port, fallen from docks, quays, and ships. The most common mishap was a broken propeller, its blades stripped off, or a damaged rudder. Sometimes the propeller shaft was bent, and the stern tube that carried it sprang a leak. All hard-worked tugs suffered general wear and tear. The long-suffering boiler was the component most often in need of repair or replacement from this cause. The *Flying Whirlwind*, twenty years old in 1895 but no great age for a tug, sprang a leak because her hull had rusted through. In a single half-year, repairs to the Alexandra fleet, from whatever cause, cost £1906 - the equivalent of about five weeks' revenue at this time. Once, three tugs were damaged in a single week. In one memorable month of 1889, three tugs were involved in five accidents, the *Alexandra* being concerned in three of them. Until then, the company had tried to discourage negligent masters by making them share the cost of the damage, or by suspending them and then reinstating them after a caution. The 'black month' persuaded the board to substitute the carrot for the stick. They set up a reserve fund, out of which a master received a half-yearly bonus of £10 (£400 today) if his tug had a clean sheet, and his engineer £5. This led to a marked reduction in accidents for about eighteen months, after which their numbers crept up again. Other operators, including Watkins, paid their masters similar bonuses.

There was always, of course, an ultimate deterrent. When the *Flying Breeze* and *Flying Whirlwind* stove in one another's plates at the end of 1888, both masters were sacked; as were the master and engineer of the *Alexandra* a year later, when she was badly damaged in collision with a ship. Two masters - including another from the blighted *Alexandra* - were dismissed during these years for 'drunkenness' - unacceptable behaviour in a company whose chairman was a temperance campaigner - or for 'fighting'.

Mersey tugmen had a reputation for keeping clear of the industrial disputes of other port workers. The only strikes that had disturbed Alexandra did so indirectly: those in the Ayr shipyard of S. McKnight, that held up delivery of the new *Hornby* in autumn 1889. Then in early March 1890 the crews of several tugs struck briefly against working with non-union ships' crews. In December that year, more than 70 per cent of the 'hitherto contented and well-conducted' tugmen joined the Sailors' Union, forming a Tug and

'Flying Breeze' which came to an untimely end in 1902 when she was crushed by the steamer 'Winkfield'.

Ferryboat branch. The occasion was an offer by the union of a specially low subscription to its friendly society, or sickness insurance scheme. The tugowners at once offered the men a friendly society of their own; to no avail. It caused further dismay in the shipping world when tug officers - masters and engineers - were seen to join, as well as men, which was unheard-of in the merchant service. The reason was, perhaps, their close affinity. All officers were deckhands or firemen made good, all came from the same working-class background, and all were thrown together as a small team working in a cramped environment, where there was no room for the hierarchies of big-ship society to grow.

The response was prompt. The Liverpool Steam Tug Owners' Association was the tugowners' trade association, to which Alexandra had belonged for the past eight months. In order to discourage the strikes without notice favoured by the New Unionism, most of the association's members demanded that their crews accept a new form of contract. In this, 2s a week would be kept back out of each man's wages until the equivalent of a week's money had accumulated; the money was to be held, with interest, against the possibility that, with a 'lightning strike', they might break the standard undertaking to give a week's notice of leaving work. The men who refused to agree these terms were sacked with a week's notice; and 600 tugmen immediately went on strike, demanding their reinstatement. The legality of the new contract was debatable, to say the least, but the tugmen made no attempt to test it.

Meanwhile some tug operators, including Alexandra, gave the masters and engineers who had joined the Sailors' Union an ultimatum: resign from the union, or lose their jobs. Alexandra's officers gave in, as did most others. The strike was still going on in mid-February 1891, but it was dying by degrees. It was never 'solid'; and 300 unemployed but experienced seamen had been brought in from other ports to man the idle tugs. It had not been a good moment to strike. Trade was unseasonably slack, ships were laid up, and, the tugowners claimed, it was no hardship for them to lay up thirty-eight of their own boats too. Like the dockers, many of the tugmen never got their jobs back, for the incomers were easily trained to replace them.

Partly, no doubt, because of business lost during the strike, the first nine months of 1891 saw a drop in income for Alexandra of perhaps two weeks' revenue over the same period the year before. This was unwelcome news for a business that had to battle for its success. In good times or bad, shipowners defaulted on towing fees: the company's solicitor was chasing bad payers in the boom year of 1890 and in the slack year of 1895. The *Hornby* towed a disabled steamer from Lisbon to Liverpool - a rare overseas job - and more than a year later her owners had still failed to pay the £266 bill. Shipowners constantly demanded rate reductions, which Alexandra usually conceded. Fighting back, tugowners kept rates up by trying to present a united front on the matter, and by eliminating competition. In May 1890 the operators in the Liverpool Steam Tug Owners' Association agreed a scale of reduced rates. Cock Tugs agreed with Alexandra to pull out of the north end docks, so long as they could still tow their associated Allan liners to dry dock. (It is significant, incidentally, that as early as this, an established and powerful tug owner was being compelled to compromise with an upstart.) But necessity overrode the agreement; Alexandra were forced independently to cut their rates to White Star, Leyland, Dominion, Warren, and other lines.

Board-room dramas added to the stresses and strains on the company. Early in 1893 George Bell Cowl resigned from the board, was given the job of Manager, and sold his shares. In February 1894, having 'forfeited the confidence of the board', he was sacked amid hints of financial malpractice. There was talk of a prosecution, but this idea seems to have been dropped, perhaps in return for his undertaking not to engage in Mersey towage again. Three weeks later the board received a petition from eleven shareholders, among them Frederic Staveley and William and Isaac Mack, asking for an extraordinary general meeting to unseat Charles Kerr, the book-keeper, who was described enigmatically as 'unsuited to be a director'. At the meeting only a minority vote supported Kerr, who accordingly lost his seat on the board, although he remained a shareholder. Two new directors were appointed to replace the casualties: J.J. Mack's son William Durant Mack, and John Hughes, the Superintendent Engineer.

No further ructions of this kind disturbed the smooth and rapid rise of Alexandra's fortunes that followed on the back of Liverpool's own boom. If one sign of the end of the recession years were chosen, it could be the fact that from October 1896, in spite of the continuing growth of its fleet, the company had to resort regularly to hiring other operators' tugs to cope with the volume of its business. The dangerous years for Alexandra, of spending money in order to secure a long-term future, were paying off.

6

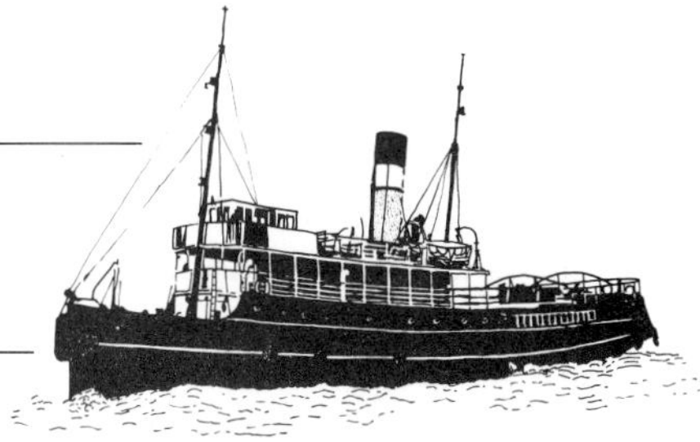

Flying Breeze 1913

Home and Abroad: 1890 - 1914

Although in the late nineteenth century Liverpool was stealing the limelight, London still led easily in sheer volume of traffic passing through the port. Here as everywhere, bigger ships of greater draught, and a demand for greater speed in turning a ship round, were putting such a strain on existing port facilities that the dock growth of the 1870s could no longer cope; in 1908 London became the last major port to have a public authority set up over it. The Port of London Authority was established to manage and modernize the river from Teddington to the sea. It was to increase dock accommodation, by building new and extending the old, and improve the Thames shipping channel for modern, big vessels by dredging. By 1914 the south bank of the river had 194 acres of wet docks and six miles of quays; the north bank down to Tilbury had upwards of 520 acres. Half the tonnage entering the port of London came to the 252 acres of the Victoria and Royal Albert Docks. Nothing available in other British ports compared with this accommodation in sheer size.

Nor was it just a matter of more of the same. Already there was dramatic proof that London was looking to the future, in the shape of bulk oil terminals at Thames Haven, where oil was discharged and stored. As a lubricant, as bitumen for road surfacing, and as kerosene for lamps, oil was already in widespread use. Now it was also the coming fuel. Motor cars and motor cycles - nearly 230,000 of them in Britain by 1914 - used it in the form of petrol; and in 1912 the Royal Navy decided to go over from coal to oil firing to make steam. The ships that brought in the oil were a new breed - tankers, the first of the bulk carriers, with their characteristic aft location of the engines to minimize fire risk from their furnaces. The first 'tank steamer' was the Tyne-built, 2300-ton *Gluckauf* of 1886, which as a further fire precaution had electric lighting, to eliminate naked flames.

As always, the London tug operators were there and expanding, to absorb new business. By 1895, John Page had bought out Thomas Elliott's share in the business bearing his name, and in 1902 set up the Elliott Steam Tug Company Limited. Dick and Page remained as the fleet's agents, and the boats were always known as 'Dick and Page' tugs. Retaining his lighter tug fleet, W. H. J. Alexander went into ship towing in 1899, when he built two tugs that he christened *Sunrise* and *Sunshine* - better known for short as *Rise* and *Shine*. Henceforth the names of all his boats incorporated the word 'sun', and the fleet became known as the 'Sun' tugs. Alexander's ship tugs became bigger and more powerful. The 130-ton steel screw tug *Sun* of 1906 began the trend, with her triple expansion engines. The 199-ton *Sun II* was the most powerful tug on the Thames when she was new. She and the *Sun III* started another fashion, for naming the Alexander tugs with numerals: but there was never a *Sun XIII*.

Gaselee, too, diversified from lighter tugs into ship towage. He bought the 170-ton *Result* and *Revenge* in 1890 to tow sailing ships out to the Nore, Hope, or Sunk light, or to the Downs, and go seeking on the way home. At the same time Gaselee enlarged his small-boat fleet in 1892 with the 54-ton *Bee* and *Mosquito*. From 1895 his tugs towed barge-loads of wood pulp and coal as far as the paper mills at Sittingbourne on the Swale, bringing paper back to the Thames. His tugs also fetched live eels from Thames Haven to Billingsgate Market. In spite of a strike by Thames watermen, which inevitably damaged a business catering largely for barge and lighter traffic, Gaselee had a fleet of eleven tugs by 1900 - including the *Lizard*, the first to be built for him - and was also building hoppers, the barges used to carry away mud from dredging operations.

Second in size to Gaselee's lighter and tug barge fleet was that of J.P. Knight. Knight, which was still independent in 1986, started with three small boats at Blackwall in 1892, under the name of the Kaiser Steam Tug Company - *Kaiser* being the name of his biggest and most powerful tug. The larger lighterage companies, such as Alexander, owned their own tugs, while smaller ones hired tugs from the likes of Knight and Gaselee. Knight shared with Alexander and Gaselee an ambition to grow into ship towage - it definitely brought more prestige - and achieved this in 1894 with his 250 hp, 69 ft sea going boat *Augusta*. But barge towing remained his bread and butter. In 1896 he began towing coke from London to the Medway cement works and lime kilns, and bringing cement, lime, and paper back to London. The metropolis, capital of business and culture, hugely expanding, was always greedy for building materials and paper. By the end of the following year Knight had eight tugs at work, six of them new-built. One of these was the 70 ft *Kestrel*, which had triple expansion engines developing 325 hp from a boiler pressure of 165 psi. A small but powerful tug of this kind cost Knight £4500.

Labour troubles in the early 1900s - coal strikes, and dock strikes on the Thames - hit the lighter tugs particularly hard because they led to a rate war among their customers the lightermen, and this in turn caused the lightermen to force down towage rates. There ceased to be enough profitable work on the Thames for both Gaselee and Knight, and Knight came to concentrate his efforts more on the Medway.

Gaselee, and also Watkins, helped in the building of Tower Bridge, completed in 1894. Gaselee's *Wasp* and Watkins's *John Bull* were on station to keep ships and barges from colliding with the bridge piers. The first merchantman to pass under the bridge when it was opened was the little old *Punch*.

The pattern of the tug had stabilized, with the development of the screw boat powerful enough to deal with any tows for which it was designed, strongly built but moderate in size for ease of handling, and economical to run. The Watkins river tug *Nubia*, of 1890, was an advanced example of the breed; a steel screw tug with 350 hp triple expansion engines and boilers working at 135 psi, yet, with a ship in tow, burning less than 2 lb of coal per horsepower per hour. Her 88 ft length made her useful for handling the P&O liners which, although becoming bigger, were still coming up the crowded Thames to the Royal Albert Dock rather than tying up at Tilbury. The 136-ton, ocean-going *Scotia*, built in 1894, was of typically modest size for her type. In the same year the grand old *Anglia* went to the breakers' yard to be converted, humiliatingly, into a lighter. Many paddlers remained in service in rivers and docks, though few new ones were being built for commercial operators. In 1902, John Masefield could still write convincingly:

> the tug has got the towrope and will take us to the Downs,
> Her paddles churn the river-wrack and all the filth of towns.

In spite of the port of London's expanding trade and docks, the years to 1914 were as difficult for Watkins as for other operators. There were expensive accidents. In 1891 alone, the *Tasmania, Oceana,* and *India* (one of the last Watkins paddlers to be built) were all sunk by steamers - the last run down by a collier without lights. All were raised to tow again; but the valuable *Oceana* was holed and sunk once more eleven years later, this time near the Goodwin Sands by her tow, the Aberdeen clipper *Orontes*. In 1908 the *Persia*, only six years old and also valuable, was sent to the bottom near the Sunk light. There were failures of reliability, too, which were out of character with the Watkins reputation. The steel screw tug *Mona* was a bad buy that broke down constantly, and was quickly sold. When the *Hibernia* - second of the name, a modern, compound engine screw tug - broke down, she had to be towed out of trouble by a sailing ship, no less.

A transport workers' strike in May and June 1912 hit Watkins as it did other owners, all of whom needed coal with which to function. The old *Cambria* and *Iona*, by now the company's last remaining paddle tugs, had to be laid up, for at a time when fuel supplies were uncertain, they were too extravagant to use.

More damaging in the long term was the growing competition for south and east coast tugs from continental operators, which until the last decade of the century had not lost anyone any sleep. But by the end of the 1880s foreign tugs were undercutting British owners' rates in the rivalry for the sailing-ship business, where low overheads mattered most. Ten years later British operators were feeling a sharper chill. The Germans, the Belgians, and the French - all the seafaring nations of the Channel and North Sea - were building modern screw tugs not just for their own river work, but for seeking and long-distance towage; and the gap between their rates and those charged by British owners was getting wider.

In long-distance towage, the most serious competition came from the Dutch; in particular the firm of L. Smit of Rotterdam. They began ocean towing in the 1890s with two 750 hp screw tugs. Smit came to be known for the towing of floating docks across the world, and succeeded in making the field their own by means of specialized skills, competitive rates, and unprecedentedly powerful tugs. By 1898 they were operating tugs of up to 1500 hp - a power output as yet unheard-of among British tug owners. Smit's most notable feat of the time was in 1908, when two of their tugs took a British-built floating dock 385 ft long by 95 ft wide from the Tyne to Callao in Peru; a distance of 10,000 miles. Not even coastal waters were safe from the Dutch. There was an outcry when, four years later, four Smit boats were hired to tow another British-built floating dock from the Tyne to the Medway.

Watkins continued to get long-distance tows, but the days were gone when they, as the major tug operator in the south and east, could expect the cream of the work. The *Hibernia* towed a steamer to the Persian Gulf in eleven days less than it took a Dutch tug towing a sister ship, and returned to Port Said without coaling - a distance of more than 3000 miles - but such exceptions tended to prove the rule. It was only the French who seemed still to be dependent on British tugs, as when in 1908 the *Oceana* was engaged to tow a French passenger liner from Bordeaux to a breakers' yard at La Spezia in northern Italy, for lack of a French tug powerful enough. Perhaps because their traditional market was contracting, Watkins decided to extend their operations to another port - the first time they had done so since their unsuccessful invasion of the Mersey in the 1850s.

Until the early 1800s the Tees had been navigable by small boats as far as Stockton - a mere dozen miles from its mouth. Below that point was an estuary two to three miles wide, full of constantly changing shallows and channels as little as three or four feet deep at low tide. The development of the export trade of the north-eastern coalfields prompted improvements in 1810; but transformation dated from after 1830, when the first house of the town of Middlesbrough was built. Iron from the nearby Cleveland Hills made Middlesbrough. Ironmasters bought 500 acres of land, on which they built a mushroom town of iron foundries and houses for the foundry workers. The Stockton & Darlington Railway arrived at Middlesbrough in the year of its foundation; as did the first shipment of coal for the foundries at its new quays.

Middlesbrough's wet dock was opened by the railway company in 1842, for colliers. But after a public body - the Tees Conservancy Commissioners - was formed in 1852, and began to make large-scale improvements to the navigation of the river, general cargoes began to flow in as well. By 1887 the Commissioners had built the North and South Gare, breakwaters at the mouth of the Tees that gave shelter to big ships anchored or berthed within. The river was widened and deepened; channels were buoyed and lit. By 1898, Middlesbrough had more than 25 acres of dock space, with a 33 ft depth of water.

The town's great boom began with the new century. In 1910, 3½ million tons of shipping passed through the port, and a year later Middlesbrough's trade was the most valuable of any port in the north-east. By 1914, Middlesbrough and Hartlepool between them formed Britain's leading iron and steel manufacturing centre. Steel called for foreign ores, imports of which soared from 1.7 million tons in 1911 to 2.6 million tons two years later. Chemical manufacture started in 1903 with the building of Brunner Mond's works. In 1910 large-scale ship-building - albeit of small ships - began on the Tees when Smith's Dock Company launched sixty-five vessels. Another hundred followed in 1911. The Conservancy Commissioners encouraged the spread of industry by reclaiming land for building from the river.

Steam towing had come to the Tees before Middlesbrough was thought of: on Christmas Eve 1824 the steamer *Albion* towed the collier *Cumberland* from the river mouth to Stockton. But the *Albion* was turned into a sailing ship ten years later; it was only in 1835 that tugs began to work continuously on the Tees. Between then and 1840, at least three companies were formed to operate them.

Twentieth-century expansion brought fierce competition among Teesside tug owners. Among them were the North Eastern Railway, who owned the docks; the Tees Conservancy Commissioners; Charles Duncan & Sons, founded in 1857; the Robinson Tug Company, which went back to 1839; William Crosthwaite's Tees Tug Company; and George Petrie. Their little tugs - none over 450 hp - worked mostly in the river, ship handling and lighter and barge towing, although they also towed up and down the coast, and went seeking in the North Sea. William Watkins of London, lured no doubt by the river's prosperity, decided to join in the hunt for business. In 1911 they joined forces with George Petrie to form Watkins Petrie & Company, sending six tugs to the Tees including the six-year-old *Hibernia*, the equally modern *Liberia* with her triple expansion engines, and the *Scotia*.

Sadly, local competition for the available custom was too strong, as it had been in Liverpool half a century earlier. In 1913, Watkins withdrew to the Thames and Watkins Petrie, together with the *Liberia,* were taken over by Crosthwaite. So, too, were Duncan's two tugs. As it happened, it was as well that Watkins extricated themselves when they did, for the Teesside bonanza ended a year later. Imports that year slumped to 1 million tons of goods.

Liverpool, meanwhile, was enjoying the peak of its prosperity, which was punctuated only by the occasional slackness in trade. The old staple imports of raw cotton and grain flooded in as never before - around 690,000 tons of cotton and 1.2 million tons of grain annually in the years up to 1914. To these were now being added new imports - chemicals, and rubber for the burgeoning bicycle and motor tyre industries. But Liverpool's exports were its real treasure - cotton goods worth £60 million in 1913 alone; iron and steel worth £25 million; woollens worth £10 million. There were new exports, too; particularly heavy engineering equipment, such as railway machinery, for newly industrialized countries - £15.1 millions-worth in 1901, £34.7 millions-worth twelve years later. Re-exports to Europe, negligible at the turn of the century, had leapt in value to £25.1 million by 1913.

The number of ships carrying the goods in and out did not change much - from just under 25,000 to just under 26,000 in 1913 - but the growth in their size was reflected in their total tonnage, up by nearly a third from 12.4 million to 18.4 million tons. The typical cargo liner of 1912 was of 7700 tons or thereabouts, which was half as big again as a transatlantic passenger liner of forty years before; and she might be nearly 500 ft long. The most dramatic increase in size was seen, as usual, in the passenger liners. The White Star liner *Celtic* of 1901, at 18,900 tons, was the first ship to exceed the *Great Eastern* of 1859 in gross tonnage. In addition to the new customary 'Grand Hotel' features, she had a gymnasium and Turkish baths. The transatlantic liner out of Liverpool culminated in the 46,000-ton White Star sister ships *Olympic* and *Titanic,* launched in 1910 and 1911, and the biggest of all before 1914, the 47,000-ton Cunarder *Aquitania,* which carried 3250 passengers and 1000 crew.

Liverpool's docks went on growing to make room for more and bigger ships. Another 84 acres of wet dock were added in the ten years before 1898, and the south docks were remodelled; but no sooner was this done than £5 million worth of extensions and improvements were put in hand. Ten years later, a further £3½ million project - the 25-acre Gladstone Dock - was begun. The first part of the dock was opened by King George V in 1913. By the following year Liverpool had sixty-five separate docks totalling over

600 acres, and more than thirty-six miles of quays. Birkenhead had 172 acres of dock, and ten miles of quays.

The departure in 1907 of four liners - *Teutonic, Majestic, Adriatic,* and a newer *Oceanic* - to inaugurate a new White Star service from Southampton to New York did not retard the boom. Liverpool reacted by setting out to attract more cargo liner business to replace the lost trade, to such effect that by 1914 the Mersey was handling 36 per cent of all the nation's exports, and around 24 per cent of its imports. There is no reference in surviving tug company records to show that they had felt a draught.

Some famous old tug businesses disappeared - Jolliffe in 1908, into the embrace of Alexandra, and the doyen of them all, the Liverpool Steam Tug Company, two years later. But other operators crowded in for a share of the trade. Russell Rea had set up as a coal merchant in Liverpool in 1872, taking his brother James Hall Rea into partnership in 1897. The Rea Towing Company, which was established on the Mersey in 1899, was an arm of what was by now the many-tentacled R. & J. H. Rea organization - coal merchants and collier, barge, and tug operators of Liverpool, London, Cardiff, and Southampton. The Rea Towing Company's tugs were small dock and river boats, intended mainly for barge and lighter work. The first was the 94-ton *Cumbria,* built for coal barge towing at Southampton in 1895 and sent to the Mersey, where she was renamed *Hallgarth.* The *Holmgarth* also arrived on the Mersey in 1899. The *Aysgarth,* in service in 1900, was a 20-ton shrimp by comparison. The next additions, of 1902 and 1903, were the 67-ton

In 1909 the 'Brunswick's' engineer put his engines to ahead instead of astern, ramming the lock gates at the Alfred Dock. He got the sack!

Fallgarth and *Edengarth*. In 1903 the Rea Transport Company was formed with two tugs to tow barges on the Manchester Ship Canal.

More shipowners were taking to running their own tugs. In 1891 the White Star line built the 619-ton, twin screw tender *Magnetic*. Ten years later Elder Dempster, who managed the Canadian Beaver line, bought three boats. In 1903 when Elder Dempster sold their Canadian interest to the Canadian Pacific Railway, who operated the transatlantic liners *Empress of Britain* and *Empress of Ireland*, their three tugs were taken over by the Mersey Towing Company, a subsidiary of the CPR.

In these years Alexandra went from strength to strength, in spite of some dropping-off of custom in 1902 and 1908-9. Their policy of spending for future good times paid off, in that although expenditure on new tugs accelerated, net profits soared, after stagnating in the years of recession. They had stuck at around £850 a half-year almost since the firm began trading; now in some years they quadrupled this. Apart from the usual dividends, at the end of 1901 the shareholders got a bonus of £1 per share. That half-year, results were regarded as 'gratifying', which was strong language for the chairman's report. The same word was used of the first six months of 1904. The shareholders usually continued to pay themselves 2s 6d a paid-up share and 1s an ordinary share - profits notwithstanding, the emphasis was still on building up reserves, in a highly capital-intensive and risky business.

Revenue grew steadily. In February 1904 weekly income exceeded £1000 for the first time, and five years later it regularly reached that figure. Receipts dropped marginally in a time of 'general depression' on the Mersey in 1902; but for Alexandra, business was 'brisk' and profit 'satisfactory', considering the conditions. In the first half of 1908, despite 'general depression', income and profits reached new heights. In the corresponding half-year of 1914, revenue was £33,637: between three and four times the figure of 1897. But the revenue was not sufficient for the company's plans. In October 1901 they increased their borrowing power to £25,000, to meet the outlay on three new tugs. There was no lack of willing lenders, including the bank. In 1903 there was a new share issue of 3,000 £5 shares, which, on being subscribed, brought the working capital up to £50,000.

Taking advantage of the new Companies Act, in 1908 Alexandra turned itself from a public into a private company. It could no longer call on the public to subscribe its shares; but the restriction the Act placed on the transfer of shares in a private company, and on the number of its shareholders, made the business easier to control. As a bonus, a private company was compelled neither to publish balance sheets nor to distribute reports to its shareholders, which improved security. Alexandra's directors owned shares in the Liverpool Steam Tug Company, enabling them to keep an eye on their rival's progress and plans.

It was a sign of good times that Alexandra was very seldom forced to cut

'Brunswick' and 'Nelson' were ordered from John Jones of Merseyside in 1899. This was a new builder for Alexandra.

its towage rates. Indeed, in late 1911 and early 1912 it was able to raise them by up to 10 per cent to some customers. Cunard accepted the new rates; White Star refused to pay them. The charge for towing from berth to river or back was now £6 for the first tug, and £5 for the second, whether a shipowner hired an Alexandra or a Cock tug. The biggest liners paid more: Cunard were asked £20 per tug for towing the 32,000-ton *Mauretania* in 1912.

Familiar difficulties cut into profits and revenue, while doing nothing to check the overall upward trend of the company's fortunes. Bad debts persisted - indeed, they became institutionalized, when a permanent bad-debt reserve was set aside. Coal prices were a regular preoccupation, particularly in the autumn, when new coaling contracts were negotiated. The rising price of coal was a new worry, that could become critical with an expanding fleet. In 1905, 13,500 tons of coal were needed, and it was expected that the figure would rise to 14,000 tons in 1906. With prices of between 10s and 11s a ton, coal alone represented an annual expense of at least £7000, or ten weeks' revenue. On the two occasions when the board had to report a reduced profit on a half-year's working, the price of coal was one reason given. (The other - the tugman's trade being a dangerous one - was insurance premiums to cover the company against claims under the Workmen's Compensation Act of 1898.) Still, it was an ill wind. From September 1908, W. N. Bicket & Company, a coal shippers' agency run by Alexander Bicket's second son, were regularly awarded one of the company's contracts, in the first instance for supplying 400 tons a month.

To absorb the expanding trade of Merseyside, Alexandra intensified its building programme, modernizing and enlarging its fleet. The four surviving original tugs were eliminated, deliberately or by accident. By 1901 the *Flying Whirlwind* was no longer worth repairing, and was sold to the breakers. The *Flying Tempest* went the same way in 1907, and the *Flying Kestrel* came to the end of her useful life two years later. The *Flying Breeze* had been crushed by the steamer *Winkfield* in December 1902 and sunk, a total loss. The ancient *Brocklebank*, by far the oldest tug in the fleet, if not an original member of it, had been sold in 1897.

To replace these boats, and to make more acquisitions, new tugs were ordered more and more frequently. The *Wellington*, a copy of the *Langton, Canada*, and *Huskisson*, was delivered by John Cran in July 1897. A new *Nelson* arrived in March 1899, but before she did so, her duplicate, the *Toxteth*, was ordered; and a month after the *Nelson* joined, two more like her were put in hand - the *Brunswick* and *Waterloo*. These four tugs, and also the four that followed, ordered between late 1900 and late 1901, came from a different builder, John Jones of Merseyside. Alexandra always put their contracts out to tender, and the lowest quote got the job. In November 1901 the company reverted to Cran for their next orders, the slightly smaller (127-ton) *Sandon* and *Coburg*. All the other additions to date were of a fairly standard pattern: steel screw tugs of around 90 ft and 150 tons, costing between £5000 and £6000.

Having added ten tugs in about three years, Alexandra allowed themselves a breathing space for eighteen months, then once again went to another builder, J. T. Eltringham of South Shields, for the *Wallasey* and *Harrington*.

'Coburg', 1902 cost her owners £600 in compensation when she collided with a coaster in fog (Albert Hayes — Salford).

'Huskisson' under way in 1914. The skipper was still at the mercy of the elements. A canvas dodger afforded little protection (Albert Hayes — Salford).

These were of standard design, as were a new *Brocklebank* and *Toxteth* ordered from Cran in April 1903. The first *Toxteth* had not lasted long in Alexandra's hands, being sold to the Admiralty in July 1901 for £8500 - £2800 more than she had cost new. A tug company that built a boat that happened to suit Admiralty requirements had no difficulty in turning a handsome profit on her. The *Clarence*, ordered in December 1903, had the highest boiler pressure of any Alexandra tug so far built - 170 psi - but a new, big tug put on the stocks in the following year was a greater departure. This was the 201-ton *Herculaneum*, 100 ft long, with a boiler pressure of 180 psi. The next three tugs, ordered before the end of 1905, reverted to type.

There was another eighteen-month break before a new tug was started - another *Alexandra*, the original having been sold. In place of the *Hornby*, wrecked off Greenock, another boat of the same name was put in hand in July 1907. These two marked a renewed trend towards bigger boats, at 168 tons. A new tug called *Egerton*, which did not last long in Alexandra hands,

In 1908, Alexandra took over the W&T Jolliffe fleet. One was the large but ageing 'Jane Jolliffe' built in 1888 (National Maritime Museum — Greenwich).

was of the same type. The *Herculaneum* had been sold, too, and another big tug of the same name was substituted in 1910. The original *Huskisson* went to Crosthwaite on the Tees; another *Huskisson* was a very modern boat of 194 tons developing (for a river tug) a formidable 1000 hp. The last new tug to be delivered to Alexandra before war broke out in August 1914 was the 211-ton *Gladstone,* named after the new dock.

The stream of newcomers were all ship and barge tugs for river and dock towing. A very different kind of boat had been put in hand in Eltringham's yard in July 1906: the big passenger tender *Herald,* 387 tons, 125 ft long, and, at £11,900, costing more than double the average river tug. With this vessel Alexandra entered seriously into competition for Liverpool's expanding liner passenger-handling market. Purpose built, a miniature liner herself, she was no mere tug that had been given a passenger certificate to enable her to nibble at the passenger trade, as were her sisters. The *Herald* was immediately successful. In 1909 she was reinforced by a new *Egerton,* a 272-ton, 105 ft boat built as a tender by Cran for nearly £7800. Two more, even bigger, tenders were ordered in 1911. The new *Flying Kestrel* was a 434-ton boat, 136 ft long, with a promenade deck for her 699 passengers that occupied all but 20 ft. The other was a new *Flying Breeze,* intended as a duplicate of the *Herald* but costing a thumping £16,315. The *Flying Kestrel* was quickly sold, and another tender of the same name (but at 700 tons getting on for twice her size), was substituted in 1913. Massive by Mersey tug standards, this was a 140 ft boat with triple expansion engines developing 1750 hp, and equipped additionally with fire and salvage pumps.

So by 1914, Alexandra had a fleet of twenty-seven modern tugs and tug-tenders, all built new for them, nearly all less than fifteen years old, and in numbers more than doubled since 1896. Yet even expansion on this scale was not enough to meet demand. In 1908, Alexandra had taken over, for a price of £25,000, the six tugs of the W. & T. Jolliffe fleet. Two were big, though ageing, ocean-going tugs: the *Sarah Jolliffe* of 333 tons and 900 hp, dating from 1890, and the 258-ton *Jane Jolliffe*, built in 1888. The *Andrew Jolliffe* was a 180-ton, 600 hp boat launched in 1894; the 163-ton *Vivian Jolliffe* dated from 1897. The 178-ton *Hannah Jolliffe* was eight years old, while the newest of the fleet was the *T.A. Jolliffe* of 1901, 650 hp and 199 tons. The *Andrew Jolliffe* was renamed *Wapping* after a Liverpool dock, in the Alexandra style; but the company kept 'Jolliffe' names for the rest until 1924, and formed W. & T. Jolliffe Ltd to run them, so keeping the goodwill of the old name. All six tugs came to Alexandra mortgaged; but the mortgages were paid off by June 1909. With the acquisition of the Jolliffe fleet the hiring of other operators' tugs almost ceased, effecting a significant economy. Little is known of the work done by the Jolliffe boats under Alexandra ownership before 1914, nor of the revenue they earned; but what is known suggests that they continued to be used for the open-sea towing for which the name was best known.

Sea towing was one of several initiatives that Alexandra exploited between the 1890s and 1914; though this one made a slow start. Tugs were released from the Mersey only when they could be spared, which was seldom: for most of 1900, there seems to have been no sea towing at all. Three years later a gradual build-up began. The *Hornby*'s tow from Holyhead to the Tuskar light, and the *Conqueror*'s from Holyhead to Glasgow, were both thought worthy of notice, though bringing in only £90 and £55 respectively. In May 1904 the company quoted for towing barges from Glasgow to Madeira. The *Hornby* was to be given a set of sails to fit her for the voyage - not something she normally needed on the Mersey - but it is not known whether the tow was carried out.

It seems that from now on Alexandra were going out of their way to look for sea work. It features in most weekly returns of revenue, occasionally with significant sums amounting to around a quarter of the total. When the work dropped off, as in November 1905, it is noticeable that 'only' the *Flying Kestrel* and *Hornby* are to go to sea, suggesting that more tugs were customarily so employed. Work included the towing of hoppers from Birkenhead to London and from Birkenhead to Plymouth. A little foreign work was being done: there were tows to Flushing in Belgium, and to La Pallice on the Bay of Biscay.

Nearer home, Mersey towing contracts were made with new customers

'Sarah Jolliffe' of 1890 was another of the W&T Jolliffe fleet taken over. Slightly younger than the 'Jane' she was heavier and had 900 h.p.

such as Houlder Bros and the Booth, McIver, and Warren lines, as well as with old friends. Alexandra won all the towage of Alfred Holt's Blue Funnel line for a year, on condition that they made no salvage claims against his ships. Now Alexandra towed Cunard liners such as the *Campania* and *Mauretania,* in addition to the company's humble coal barges. Most significant of all was Alexandra's connection, established in 1910, with Harland & Wolff of Belfast, who were building the liners *Olympic* and *Titanic* and big warships for foreign governments. On and off, up to five Alexandra tugs at a time worked across the Irish Sea for at least three years before 1914. These were highly lucrative contracts that might earn Alexandra well over £800 for two weeks' work.

In 1900 Alexandra had crossed the Mersey and established an office at the Morpeth Branch dock in Birkenhead. A steady, if unspectacular, income sprang from this enterprise, averaging around 10 per cent of the figure for Liverpool dock and river work. Manchester Ship Canal jobs never amounted to much, either: there were weeks when nothing at all came in from this quarter. After its early promise, the canal seems to have been a disappointment to Alexandra.

Salvage, too, was still a minor element in the company's revenue: incidents were dramatic, rather than frequent or profitable. In 1907 the *Herculaneum* helped to salvage the broken-off stern of the liner *Suevic* from rocks near the Lizard. When the *Mauretania* went ashore in the Mersey, no fewer than eight Alexandra tugs aided her, but got nothing at all, either because the company had a towage agreement with Cunard that excluded salvage claims or because Cunard were valued clients. The advantage could be the other way. When White Star claimed £600 for damage to the *Laurentic* by the tug *Toxteth*, they finally settled for £15.

Many of the miscellaneous jobs that came Alexandra's way were more fruitful. The *Wallasey* was chartered for no less than thirty-nine weeks at £50 a week to tow dredgers and mud hoppers at Fishguard. For the winter of 1912-13, the *Herald* was on charter in a rather different social milieu, acting as tender in Monaco to visiting liners for £535 a month. The *Flying Kestrel* tender later took part in this work, too. On the Mersey at Easter, and again from August to October, Alexandra tugs made money more consistently, if prosaically, carrying trippers on river cruises. Weekly earnings from this source could be as high as £80. And in October 1910 the Liverpool Salvage Association hired the *Flying Kestrel* for £25 a tide to take pumps and other equipment to a ship stranded in the Pentland Firth. Other odd jobs included towing dredgers in Southampton Water and two days of somewhat smarter towing at the 1911 Coronation Spithead naval review. The relative value of the different kinds of work is illustrated in a typical week's income shortly before the outbreak of war in 1914. Liverpool dock and river work brought in £1138, Birkenhead work £41, tendering £55, sea towing £50, and pleasure cruises £36 - a total of £1320.

In human terms, the price of rapid growth was high. Strain on the crews is suggested by the accidents in which they were involved - not so much by their numbers (fewer than a dozen were recorded, in seventeen years) as by their gravity. The *Conqueror* broke down off the Calf of Man in the first days of 1905, and was driven ashore in a gale. Six men of the eleven on board were drowned. Their employers paid flat sums of between £10 and £30 to the widows and mothers, and £1 or £2 to the survivors, presumably for the replacement of lost belongings. This was a terrible time for the men of Alexandra, for six weeks later the *Clarence* vanished in the Irish Sea while seeking. She had towed a sailing ship to the Tuskar light, and was last seen nearby 'in pursuit of a vessel'. On this occasion the company gave the crew's dependants pensions of between 5s and 10s a week. Less than a year later the *Sandon* went down in the Mersey with all eight on board, overrun by a steamer. The *Hornby* was sunk in a collison with a steamer on the Clyde in March 1907, though without loss of life; and both tugs were raised to tow again another day. A few years later a *Hornby* crewman was killed by a parting towrope. The third week of May 1909 was a bad one for Alexandra. The *Hornby* and *Herculaneum* collided, bringing a severe reprimand for both masters, who were absent from their boats at the time. The *Hornby* was in collision again the following week, with the *Alexandra*; but it was the *Alexandra*'s skipper who was sacked. Next, the *Brunswick*'s engineer put his engines ahead instead of astern, ramming his tug into the lock gates at the Alfred Dock. He and his mate were fired.

Alexandra established a fleet of tug/tenders to provide service for the passenger liners which called at Liverpool. One such was the 'Egerton', 1908.

After the disastrous failure of their strike twenty years before, there was no serious unrest among the crews until 1911. The new dispute came as a part of a spell of constantly erupting industrial unrest that beset Liverpool, and the nation, until war came in 1914. One cause was a rise in prices that began in 1907, matched by a drop in the purchasing power of the pound that had already started, but not by any general increase in wages. A Liverpool seamen's strike in June 1911 succeeded in winning them higher wages. It drew in the dockers, partly in sympathy and partly for more money; and then, in the first days of July, the tugmen too. Events occurred in this order because the Liverpool tugmen had severed their connection with the seamen's union, which had done nothing for them in 1891, and had joined the National Union of Dock Labourers instead. It was now this union that spoke for them, as the seamen never had.

The seamen went back to work on July 7 pending negotiations; then came out again. This time they stayed out until the end of August; although there is no record of their getting a rise in wages, or any other benefit. Since it involved all the port workers, the dispute had a pronounced, if temporary, effect on Alexandra's revenue. In the first week of the stoppage it fell by nearly half, in the second by four-fifths, and in the third to a miserable £100. Other disputes could work to the company's advantage, so long as crews did not unite. In July and August 1912 there was a strike at Birkenhead that did not involve Alexandra tugs, and two of them made a great deal of money there while on hire.

The stresses were hard on managers as well as men. Three out of the four survivors of the original board died in harness, in the space of little more than three years. Thomas Durrant died at sixty-six in November 1899, to be replaced by the forty-six-year old Alexander Bicket in the first weeks of the new century. William D. Mack followed Durrant to the grave less than a year later, and was succeeded on the board by his father, James Jermyn Mack, who was the same age as Bicket. William Poulsom was carried off in January 1903, at the age of seventy-two, being replaced as chairman by the engineer John Hughes. Shortly afterwards William Poulsom's son Frederick A. J. Poulsom joined the board. J.J. Mack became Deputy Chairman, and in January 1909 was elected Chairman on Hughes retiring from the post. Hughes remained on the board. By 1903 only Frederic Staveley survived of the six founding directors of 1887.

Between 1890 and 1914 the customers - the ships that the tugs towed - underwent a transformation. We have seen that as ships became ever bigger, in search of economies of scale, they demanded tugs that were both powerful and handy: when moving the latest passenger liners, tugs needed to be numerous, too. The ship's sides were like cliff walls, catching every puff of wind. Three, four, or more tugs might be needed, to shift or hold them against wind and tide. When the 52,000-ton *Leviathan* (formerly the Hamburg-Amerika liner *Vaterland*) ran aground in the Mersey late in 1918, she needed thirteen of Alexandra's tugs to pull her off. It may have been at this time that tugs began to push as well as pull their charges, mounting bow fenders for the purpose. Inertia, wind resistance, deep draught, and other factors generally made a big tow respond sluggishly and, once moving, likely to run out of control. In a tricky or delicate situation, or in a confined space, a ship's movements could be 'fine-tuned' by applying pressure immediately and precisely at any point along its hull.

Ships began to change in their mechanical specifications also. The first motor ships appeared - the internal combustion engine afloat. A diesel engine had been used to propel a canal barge in France in 1902; but the successful application of the diesel engine to ocean-going ships happened in Denmark, where in 1912 the 7400-ton *Selandia* was built. It was lighter and - lacking a boiler - more compact than the steam engine, allowing more space for paying cargo. At the same time, the steam engine had reached the peak of its efficiency with the introduction of the steam turbine. The Parsons turbine had been demonstrated, sensationally, by the 30-knot steam yacht *Turbina* at the Spithead naval review of 1897. It, too, was lighter than a conventional reciprocating steam engine, giving a better power-to-weight ratio; and it used less coal. It was adopted first by the Royal Navy, and then, after its perfection in 1901, by the merchant fleets of the world.

Since the turbine allowed a bigger ship to be propelled at the same speed by a smaller engine, or faster by an engine of the same size, it showed itself most economically in big, fast ships - another factor that tended to increase the size of a ship and the speed of its turn-round, making more work for tugs. Furthermore, the turbine worked smoothly, without vibration, imposing less stress on moving parts, increasing reliability, and also commending itself to liner passengers. The Allan liners *Victorian* and *Virginian* were the pioneers; after the Cunarder *Carmania* of 1905, the turbine became established on the transatlantic route.

The high speed of rotation of the turbine, and hence of the propeller, was an advantage when high ship speed was the main object, as in transatlantic liners; but it was unsuitable in cargo ships, that operated more economically at lower speeds. The best of both worlds was obtained with the geared turbine, introduced in the *Vespasian* in 1910, which reduced the speed of the propeller. The turbine was by then the most highly developed technology for the big passenger or cargo liner, while the diesel engine showed itself to the best advantage in smaller vessels.

Self-evidently, the steam turbine was of no direct value to the little tug; nor, initially, was the diesel engine. As always, entrenched technology - in this case steam - had the advantage. Diesel engines needed specialized, skilled knowledge that steam engineers lacked; and coal was cheap and plentiful, at a time when oil was neither.

A diesel tug had one apparent advantage over steam: it should have been economical, because it did not need the steam tug's large crew of stokers and trimmers below decks. Furthermore, a tug spent much of its time on standby, awaiting orders to sail. To keep steam up, ready for instant departure, a steam tug had to burn coal all the time, whereas a diesel tug did not burn fuel until its engines were started; and it could then sail at once. But this factor would only tell when oil became cheap, and when the diesel engine's power and flexibility grew. Early diesels lacked power. Also a tug manoeuvring a ship in the confined space of a dock or river - its main area of operation - needed to be able to change direction and propeller speed frequently and quickly. The steam engine could do this, and could produce high power at low engine speeds. The diesel needed high engine revolutions to produce power, and could not respond so promptly. Not surprisingly, in the tug world the diesel engine was taken up first by the operators of barge tugs, which spent most of their time towing relatively modest loads forwards at a constant speed. They had appeared on the Thames in small numbers before 1914.

The tug's design changed little between 1890 and 1914, but the work of the sea-going tug was revolutionized. The demise of the sailing ship after

'Egerton' approaching Princes Stage with the Swedish liner 'Drottningholm' carrying repatriated prisoners at the end of World War II.

'Wallasey', built by J.T. Eltringham of South Shields in 1903, survived until 1959 when she was broken up. (A. Duncan, Gravesend).

about 1890 was rapid. The cost of carrying a ton of manufactured cotton goods to China in 1866 under sail had been £6. By 1913, under steam, it was £2. There was no gainsaying the logic of those figures. By 1898 there were about 8500 sailing ships, totalling 3.1 million tons, left under the British flag: fewer than half the numbers of the mid-1870s, and little more than half the tonnage. In 1913 a record 1.2 million tons of new ships left British yards; of that, just 30,000 tons were sailing ships. Six years later there were only 3 million tons of sail registered in the entire world - less than there had been in Britain alone in 1885.

Seeking became extinct as a way of life, if not as an occasional source of work. In 1898, only eight years after buying the *Revenge* and *Result* for seeking, Charles Gaselee sold them both: the customers were simply not there any more in sufficient numbers. A Watkins tug - the *Arcadia* - went seeking for the very last time a week before the outbreak of the First World War. That war was, in fact, the final nail in the coffin of seeking. The narrow seas between Britain and the Continent were so dangerous, particularly from U-boats, that the cost of a short tow from (say) Dunkirk to London might be as much as £200. No sailing-ship owner could afford that sort of money now. Surviving sailing-ship operators took their vessels off regular routes, where they could not compete with cargo liners, and put them on to tramping. This alone would have discouraged the seeking tug, accustomed to looking for a sail at a particular time and place. But tramp sailing ships now put into Queenstown in Southern Ireland or Falmouth in Devon to find out where to discharge. The shipowners, knowing where they were, contracted with tug operators to wait for their vessel, for example in the Downs or at Gravesend, eliminating the aspect of chance.

There was another way in which finding work became less hit-and-miss. Getting news of a job to a tug operator's or agent's office, and thence to his boats, before a rival heard about it had been a matter of sending a runner or a rider, or (if a tug were offshore) wagging signal flags. Unsurprisingly, tug owners were among the first to appreciate the value of the telephone, which The Alexandra Towing Company had decided to instal in its dockside office by April 1889. It was, of course, still necessary for a tugmaster to land to get his orders. The German ocean-going salvage tug *Roland* was equipped with wireless in 1903, but it would be a long time before it would be in general use. Meanwhile, when the Hamburg-Amerika liner *Steinhoff* went ashore in fog in Start Bay in 1894, the master of the Watkins tug *Hibernia*, at Dartmouth nearby, heard about it in the old way, from a man on horseback.

Epic tows began to grow fewer, as the reliability of steamers increased and dry dock facilities worldwide improved - largely on account of the efforts of the same tugs in towing floating docks around the world. These were,

Built in 1913 'Flying Breeze' is seen here with the Crown Prince of Japan on board. 'Flying Breeze' later attended the launch of both the 'Queen Mary' and the 'Queen Elizabeth'.

indeed, the new kind of tow; not a ship, but a massive, inanimate object that needed to be floated from the place where it was built to the place where it was to be used. The oil or gas rig is the latest in the line. One 5000-ton dock, destined for Havana in Cuba, was so big that Watkins, who were given the job in 1897, had to charter the New Zealand Shipping Company's 4200-ton steamer *Ruapehu* for the work. During the 6500-mile voyage she was supported by the *Oceana,* Watkins's biggest and most powerful tug. The 120-fathom towrope had to be specially made, and was 22 inches in circumference. Eight years later the Liverpool Steam Tug Company's big 380-ton *Cruizer* and 283-ton *Blazer* towed a floating dock 4500 miles to Nigeria.

Another job that had nothing to do with rescue incidentally set a new distance record. In 1905 the *Oceana* towed the oil storage vessel *Tancarville* from Portland to Sumatra - 8200 miles, accomplished in forty-five days' steaming. In the same class, and much against Watkin's better judgement, was the *Oceana's* tow of a pontoon-borne grain elevator across the Atlantic from Barrow-in-Furness to Montreal. Her owners' misgivings were justified. In mid-ocean the towrope broke, fouling her propeller. She and her tow drifted apart. The tug hunted the pontoon for twelve days, to no avail. It was eventually picked up by two trawlers near Rockall and taken into Stornoway in the Hebrides. Then the *Oceana* tried again, and this time completed the job successfully.

One heroic ship tow was undertaken in 1895, when the *Blazer* (second of the name from the Liverpool company) took the 2400-ton steamer *Colombo,* which had broken a propeller shaft, from Rio di Janeiro to Genoa - a trip of 6000 miles, which was then a record. A shorter, but because of the size of the ship, perhaps more difficult tow was successfully accomplished seven years later, after the Cunarder *Etruria,* 8120 tons, broke a propeller shaft in mid-Atlantic and had to be rescued from Fayal in the Azores. The Liverpool Steam Tug Company's *Pathfinder,* together with Cock Tugs' 254-ton *Blackcock,* H.J. Ward of Liverpool's *British King,* and a Portuguese tug towed her home to Liverpool in seventeen days.

Salvage tended to become more dramatic, and sometimes more valuable, with the growing size of the casualties. The 10,500-ton Inman liner *City of Paris* was en route from New York to Liverpool in March 1890 when one of her propeller shafts broke, wrecking the engine and flooding an engine room. The *Pathfinder* and three other tugs towed her into Queenstown on the south coast of Ireland, and thence - after pumping her dry - to Liverpool. In February 1894 the 2700-ton steamer *Loch Maree,* with a cargo of raw cotton, ran out of fuel and was abandoned by her crew. She went ashore on the Irish coast, where she was found by the *William Jolliffe* and brought into Belfast, winning salvage of £7500 for her owners and crew.

From 1892, such proceeds of salvage began to come a tug's way by a new route. Salvor and salved continued to resort to the Admiralty or county

In 1888, 'Blackcock', then two years old, is credited with towing the 4,200 ton 'Norham Castle' from St. Helena to London in 39 days, coaling only once at Goree.

courts if they wished; but they came to rely increasingly on the new Lloyd's Standard Form of salvage agreement, drawn up and signed before the salvage attempt started. The principle of 'no cure, no pay' became formalized. The gradual adoption of the Lloyd's Standard Form took most of the heat and uncertainty out of salvage claims. If there was a dispute, a Lloyd's arbitrator, whose participation was obligatory on both parties, was there to settle it. Either party could thereafter appeal to the Admiralty Court, but in practice has seldom done so.

7

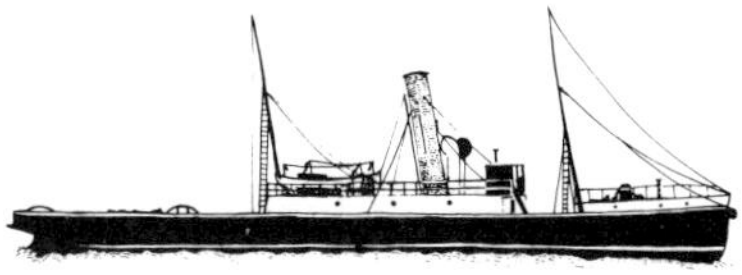

Blackcock 1886

War Service: I

On the declaration of war by Britain on Germany and her allies on 4 August, 1914, hundreds of civilian tugs were commandeered from their owners as naval or military auxiliaries, for periods ranging from one or two weeks to the duration of hostilities. Payment for their services was made by the day or month. In all, over 400 tugs were taken over at various times, more than 300 of them by the Navy.

At first there was confusion. Every single Thames tug was requisitioned, until John Watkins, the current head of William Watkins, pointed out to Whitehall that without tugs, the river would be blocked within twenty-four hours with ships carrying food and other essential supplies to Britain, all helpless, unable to move. In Liverpool a smaller proportion of tugs was taken. No more than a third of Alexandra's fleet was under requisition at any one time, allowing the rest to get on with moving ships in and out of the Mersey. Liverpool became a funnel through which troops, war material, and food poured - the busiest in all Britain. This did not mean that shipping movements in the Mersey increased. Indeed, they declined sharply under the stresses of the German mine and submarine blockade of Britain's ports. Whereas in 1914, 15,000 ships aggregating 15 million tons passed in and out of the Mersey - itself a massive drop on 1913 - in 1917 the figures had sunk to 9832 ships totalling 11 million tons, and in 1918 to 7248 ships and 10 million tons.

The commandered tugs sailed under the White Ensign instead of the Red, with Royal Navy or Royal Naval Reserve officers in charge, and under new names: for instance, the Watkins tug *Hibernia*, and later the *Java*, were inelegantly rechristened HMS *Carcass* for the duration of their service. They retained their civilian crews, with the addition of one or two naval ratings or Royal Marines to act as signallers and wireless operators. These were necessary because British tugmen had no experience of radio, with which their boats were now equipped for the first time. It was said, too, that the crews found it hard to adapt to naval discipline, after the freewheeling ways to which they were accustomed.

The requisitioned tugs towed supply vessels, troop transports, and warships, in British ports and across the Channel in France and Belgium. They also served in the boarding flotillas in the English Channel, in the narrow seas, and outside major ports, where their task was to stop merchant vessels. The purpose was twofold: to back up Britain's blockade of Germany and to help guard British ports. Each ship stopped in the Channel was boarded by a naval officer and 'examined'. If she was carrying no contraband - among which were counted Germans trying to return home to join up - she was released, and if on her way to the Thames, a safe passage through the minefields was pointed out to her. If the officer was not satisfied, he signalled to one of the armed 'mother' ships on permanent station in the Downs to send across an armed guard or search party. Most tugs were not themselves armed. Those taken by the Admiralty were the bigger tugs and tenders, at home in the narrow seas around Britain, or even in the open ocean. Strongly built, they could stay at sea in all weathers.

In company with the *Herald*, Cock tugs, and one of Frank Batey's tugs from Tyneside, Alexandra's *Flying Breeze* found herself off Milford Haven, 'examining' vessels approaching the Pembroke Dock naval base. Thirty or forty might try to enter in a single day - coasters, oil tankers, Belgian trawlers with families on board, fleeing from occupied Belgium much as Vietnamese boat people fled sixty years later. Nearly all stopped at a signal from the examining boat. If one failed to do so, a signal to the battery covering the harbour mouth brought a shot across her bows. The tug crews were strained and generally subdued, obsessed - particularly in fog - by imagined sightings of U-boats and torpedoes, which invariably turned out to be porpoises. But they had their moments of relaxation, as when a shore party came back with baskets full of blackberries from the local lanes.

Then the *Flying Breeze* and *Flying Kestrel* were sent to Stromness on Scapa Flow. There the *Breeze* and other tugs spent years as tenders to the great fleet riding in the anchorage - thirty battleships and battle cruisers; innumerable other warships, real and dummy; cargo ships, tankers, colliers, hospital ships. They took out fresh water and stores, ferried mail and personnel to and fro, and picked drunks out of the water. There were frequent U-boat scares. Once, without success, the *Breeze* joined in the search for the crew of a submarine rammed by a minesweeper. In June 1916 the *Kestrel* searched the ocean for survivors from the cruiser *Hampshire* which had been carrying Lord Kitchener to Russia when she was mined off the Orkneys. All the tug found was seventy-two bodies. Almost exactly three years later, on 21 June, 1919, the surrendered German fleet scuttled itself at Scapa Flow just as the *Kestrel* sailed through its ranks with a party of sightseeing school-children, who had a grandstand view.

Seen here alongside 'Kaiser Fredriech' at the Spithead review in 1901 is R.F.A. 'Alligator'. Built at Cammell Lairds in 1885 as 'Stormcock' she was sold to the Admiralty in 1896 (P.A. Vicary, Cromer).

The auxiliary tugs had another task, which they shared with their sisters left in civilian service. The Germans mined the approaches to Britain's ports; their U-boats were busy; and almost all the night navigation marks - lighthouses and lightships - around the coast were extinguished. There was therefore a huge increase in shipping casualties, mined, torpedoed, or wrecked. Tugs were stationed from Portland to Stornoway and in Irish ports, with the task of assisting them; although the crews and owners got little salvage, their boats being on government service. Off the north-east coast the Crosthwaite tug *Triumph* found some pioneer air-sea rescue work when she saved the pilot of a wrecked seaplane.

Owners had difficulty in meeting both the Admiralty's demands for tugs and the needs of the traffic in and out of their ports. To keep their Thames fleet up to strength, Watkins chartered some tugs, and bought a Dutch ocean-going boat, the 410-ton *Ocean*. Built originally for Smit, and now renamed *Racia*, she had an adventurous career in Watkins hands. With other tugs from Britain, she sailed around North Cape to Archangel in the White Sea, to work with Britain's Russian allies. She carried Russian VIPs home, and a cargo of precious platinum for their industries. When the Royal Mail liner *Arlanza*, serving as an armed merchant cruiser, was mined in the northern seas, the *Racia* helped other Watkins tugs to tow her to Harland & Wolff's yards at Belfast for repair. Liverpool tugs, too, took war supplies to Archangel. The *Blackcock* and *Sarah Jolliffe* towed barges all the way from Cadiz in Spain.

For the most part, Charles Gaselee's tugs stayed nearer home, towing ammunition barges on the Thames. An exception was the *Apar*, with 500 hp one of his more powerful boats, which worked in Le Havre and Dieppe. Alexandra tugs, too, served in the French Channel ports. When the Cunard liner *Lusitania* was torpedoed off southern Ireland in 1915, the Elliott tug *Warrior* out of Queenstown was one of the first vessels to reach the scene, saving seventy-four lives. The Cock tug *Stormcock* was there to help as well.

The requisitioned tugs went much further afield. In 1915, John Batey's tugs, among many others, sailed from their native Tyne to the Dardanelles, with the Allied forces that attempted to seize the Bosphorus and capture Constantinople. A Star Tug Company boat from Liverpool supported the Gallipoli landings, and the Watkins tug *Hibernia* was at the second landing at Suvla Bay, frequently under fire. She was one of the last ships to leave Anzac Bay at the evacuation; then sailed for the Persian Gulf to support the next British attack on the Turks. The greatest adventure of British tugs in the First World War came their way in Africa. Having sunk the liner *City of Winchester* in the Indian Ocean, and the British cruiser *Pegasus* off Zanzibar, the German raider *Konigsberg* sought refuge in the Rufiji river delta in German East Africa. She was inaccessible to conventional big warships, so the shallow-draught, big-gun monitors *Mersey, Humber*, and *Severn* - inshore bombardment vessels - were sent out from Britain in tow of Elliott's *Revenger*, Cock Tugs' *Blackcock*, and Alexandra's *T.A. Jolliffe* and *Sarah Jolliffe*. After the raider had been destroyed, the tugs towed the monitors back to the Mediterranean.

Home waters were much more dangerous for tugs than distant, for mines and U-boats were more dense in the narrow seas. Here, too, the hurried, hidden comings and goings of multitudes of darkened ships held special hazards for the small and vulnerable tug. Among the most dangerous of all waters for tugs was the North Sea, facing Germany. In January 1917 the Crosthwaite tug *Ida Duncan* - still manned by the Duncan family - hit a mine while towing barges. All six of her crew went down with her, including two Duncans. Later in the year the same owner's *Hero* was mined off Sunderland, and her master killed. Watkins's old *Guiana* was sent to the bottom by a British destroyer that rammed her, losing her master and three of her crew.

Sometimes a tug fought back successfully. When John Batey's Tyneside boat *Homer*, towing a sailing ship in the English Channel, was set upon by a U-boat, her Captain Gibson cast off his tow and stormed in through intense machine-gun fire, intent on ramming. He missed the submarine by a whisker, and after firing a torpedo that missed, the enemy finally gave up. At other times, a tug at war found itself in conditions it was never designed

The ill-fated 'Lusitania' with the Cunard tender 'Skirmisher' in attendance entering dry-dock in July 1907. 'Skirmisher' was in fact the first twin-screw 'Cunarder', built in 1884.

to withstand. Stranded in the White Sea in 1918, the *Blackcock* was trapped by ice and crushed.

In war conditions, a tugman might find himself in more peril from his own side than from any enemy, human or elemental. William Simmons, not yet eighteen, was second officer on board the little wooden Admiralty tug *HS67* at Greenwich in January 1918, when she sailed with a barge in tow for the Firth of Forth. In command, if not in charge, was a naval sub-lieutenant. When the *HS67* was steaming past St Katharine's Dock, firmly on course for Windsor, Oxford, and points west, young Simmons felt obliged to tell his captain that the sea was the other way. The mate, an old Watkins hand, had been unwilling to do so. Later, anchored in fog under the cliffs of the Northumberland coast, the tug's anchor dragged. Her engines were started, just in time to strip the blades from the propeller as she ran on to the rocks. The barge was holed and filled, dragging the tug on to her side. Simmons found himself crawling back along the tilted deckhouse with an axe, to cut the towrope and right the tug. The mate fired a distress rocket, setting fire to the bridge in the process. Since the tug was being pounded to splinters, Simmons then rowed the entire crew, including the ship's dog, ashore in four trips, through pitch darkness and mountainous seas.

Liverpool was out of range of the German bombers of 1914-18, so the city did not suffer physically from the conflict. In the Alexandra records there is little direct mention of the war, though plenty about requisitioning. Within a fortnight of war breaking out the *Sarah Jolliffe* had been requisitioned by the Admiralty as an examining vessel, and the *Hannah* and *T.A. Jolliffe* as tugs. By early September the *Herald* had followed them into the Navy; then the *Flying Kestrel* and *Flying Breeze*. In October they were joined on active service by the *Herculaneum, Harrington, Wallasey, Alexandra,* and *Hornby,* which sailed for the English Channel on examining duties. Alexandra tugs were scattered around the coast from the Downs to Scapa Flow, moving from station to station and task to task as the Admiralty required.

By November the Mersey Docks and Harbour Board were already complaining to Alexandra about the shortage of tugs for ordinary towing on the river. Some boats were released, but others were taken. By summer 1915 the *Wapping* had been sent to the Mediterranean to be shot at by the Turks, and by spring 1916 she was in the Persian Gulf. Several tugs, such as the *Egerton,* were requisitioned, released, then commandeered again. At any one time, between six and ten boats - a third of the fleet - were away from the Mersey on national service. For Alexander Bicket and the company's Manager, Silvester James Procter, negotiating requisitions and terms of hiring with the Admiralty, or later with the Ministry of Shipping, was a full-time job. Fortunately for the business, the tugmen who stayed in civilian work were in a reserved occupation, so the company did not lose crews as well as boats.

New tugs, in particular, were Admiralty favourites; others were sold to Allies worse off for tugs than the British or, like the *Hornby*, were chartered to them. A new *Gladstone* - replacing the first, sold to Russia - was launched in spring 1915, but she and another addition to the fleet, the *Bramley Moore*, were at once sold to the French government at a handsome profit. Two boats were immediately ordered from Cran in their place, but no sooner were the replacement *Gladstone* and *Bramley Moore* launched a year later than both were instantly requisitioned. A new *Alexandra*, built to replace her predecessor, wrecked off Orkney in October 1915, lasted rather longer in her owner's hands. In service by March 1917, she was requisitioned in December. Alexandra seems to have been able to hang on to only one new tug - the *William Poulsom*, ordered at the same time as the new *Alexandra*. All were big tugs on the lines of the original *Gladstone*, of up to 219 tons.

Compounding the shortage of tugs was the fact that the shipyards building them, forced to give priority to Admiralty orders, took a very long time to deliver. Instead of taking seven to nine months, as in peacetime, John Cran were a year building the second *Gladstone*, and eleven months with the new *Bramley Moore* and *Alexandra*. In January 1916, Alexandra bought second-hand for the first time since their early expansion - the former Dundalk harbour tug *Liberator*: she was at once requisitioned. Since any one yard, such as Cran's, was likely to be overwhelmed by Navy orders, Alexandra spread their custom wider than usual. The 25-ton *Stalbridge* was ordered from Scott & Son of Bowling on the Clyde, and the 95-ton *Sloyne*, named after the Mersey anchorage, from Abdela & Michel of nearby Queensferry on the Dee. The *Sloyne* was another tug that took a year to deliver. In January 1917, Alexandra bought a tug they had already found building at Thomas McLaren's Montrose yard - the little *Salthouse*. By the end of the year she, too, had been requisitioned. These four tugs were all well under 100 tons; the *Salthouse* was of only 63 tons. Perhaps in an attempt to gain tighter control over the vital supply of tugs, in May 1917 Alexandra bought a quarter-interest in the shipyard that still built most of their boats, John Cran & Company, for £10,000.

The shortage of tugs was aggravated by losses and accidents - and may have helped to cause them too, since what tugs there were had to work longer and harder without time off for repairs or overhauls. As late as January 1916, neither had been carried out on any boats since the outbreak of war. Such pressures must have told on their crews as well. The *Sandon*, sunk in the Irish Sea in collision with a steamer in September 1914, was raised and renamed *Huskisson*; but all her crew had been lost. Luckily, when the *Alexandra* was lost she killed nobody, and the Admiralty eventually disgorged £5000 in compensation. In April 1918 the *Canning* was sunk in the docks while towing, although she, too, was raised.

The host of lesser mishaps probably cost the company more than the few disasters - nearly sixty separate accidents were recorded during the war years, some of them expensive not only in working time lost, but in money too. Generally it was a matter of broken propellers and dented flats, as usual; but when the *Coburg* collided with a coaster in fog, she cost her owners £600 in compensation. While towing the *Aquitania*, the *Herculaneum* lost her funnel and mast in spectacular fashion. In the winter of 1918 the *Alexandra* ran on the rocks at Abbey Head while towing in the Solway Firth; and when her tow sank, a crewman of the *Herculaneum* who was on board her drowned.

From 1915, with the war, prices rocketed for everyone, including the tug owners. Furthermore, the bank rate had gone up in 1914 to 10 per cent, making loans more expensive. A new boat, that had cost £5000 to £6000 before the war, might now, like each of the two ordered in July 1915, set its owner back £10,250. The price of coal soared. By July 1916, W.N. Bicket were selling to Alexandra at 26s a ton; less than a year later, the price could be as high as 31s. As a consequence, the company's coal bill doubled during the war years.

As maritime perils grew, insurance premiums against war hazards became a major item of expenditure. Unheard-of sums, such as £9000 in a half-year, compared with £1,400 immediately pre-war, had regularly to be set aside out of profits to cover it.

As prices went up, the purchasing power of the pound slumped. Unrest among the crews, and wage demands in particular, kept pace. In 1913 the practice of giving a crewman a twenty-four hour break for every Sunday they worked had been introduced in Liverpool. By November 1914 crews were not getting their break because of the pressure of war work, so with the backing of the dockers' union they went on strike for three days, demanding extra hands on board the tugs to allow each man a rest day. The company's revenue that week fell by nearly three-fifths. The crews got their extra men. Between then and the war's end the tugmen won whenever there was a confrontation, gaining - as members of the dockers' union - wage increases whenever the dockers did so. In 1915 officers' commission was raised to 3 per cent, as a temporary 'war bonus'. The union and the tug officers asked for a bigger share of salvage proceeds, and more pay for working out to sea. They got 20 per cent of all salvage, instead of around 15 per cent; and 15 per cent of income from salvage paid for by the tide. To keep going, it was clearly worth while to the tug owners to give way on issues of money, but the result was a wage bill that went up by nearly 120 per cent in the war years. To pay for all their extra outgoings, both Alexandra and Cock Tugs put up Mersey towing rates several times - or rather, they promulgated the new rates, and then negotiated with individual customers, in the time-honoured manner. In April 1915 rates went up 20 per cent, and at the end of 1916 another 10 per cent, and by 20 per cent again in November 1917 and June 1918.

'Bramley Moore' Built in 1915, was immediately sold to the French for a handsome profit. She was renamed by them 'Courageux'.

The tugmen's militancy was probably encouraged by the undoubted fact that Alexandra - and the shipping industry in general - did unprecedentedly well in the war years, in spite of their manifold difficulties. The tugowners had much in common with the big-ship owners, who took dreadful losses - 7.8 million tons, or 38 per cent of their 1914 tonnage, sunk in four years of war - but with what was left, and was built new, made huge profits from increased freight rates and paid unheard-of dividends. For instance, the cost of carrying coal from Britain to Egypt multiplied thirty-fold between 1914 and 1918.

The Alexandra report presented to the shareholders in January 1915 spoke of 'considerable difficulty and anxiety' due to the war; but gloom evaporates thereafter. It was not in the everyday work that the real boost came, although by any normal standards the news was good enough from that quarter: in the first year of war, dock and river towing brought in around £1300 a week, and by the summer of 1918, after climbing steadily through the intervening years, the figure was well over £2000. We are now told more about dock and river revenue from the Jolliffe boats, because from 1916, Alexandra took over their management from the Jolliffe company; but it was not substantial: in one four-week spell, it averaged £32 a week. The rise in revenue was achieved in spite of the decline in shipping movements on Merseyside, which suggests that the rises in towing charges did better than merely compensate for huge additional expenses.

Old friends continued to bring Alexandra their towing. Harland & Wolff engaged between two and five tugs half a dozen times a year, for launches or for towing damaged ships to Belfast. New friends were made. One contract gave the company all Elder Dempster's towing for a year, even though they had their own tug company. Less surprisingly, perhaps, Alexandra got a towage contract from the United Grain Elevator Company, a successor to Poulsom's Grain Elevator Company. Until his death at the age of seventy-one in April 1915 - which removed the last of Alexandra's 'founding fathers' - Frederic Staveley had been that company's chairman. J.J. Mack replaced him in the role.

Salvage, naturally enough, was a source of increasing revenue. Mines and torpedoes, as well as natural hazards made worse by congestion, haste, and darkened ships, produced eight or ten times as much work, but there had been so little before the war that this still did not amount to very much. It was rare indeed to be called to the aid of such a ship as the *Empress of Britain*, as three Alexandra and two other tugs were in September 1918. The Alexandra tugs shared a sensational £37,500 between them, which was a windfall for their crews, now on 20 per cent of revenue. On average, each man must have received around £300.

Sea tows, by the tugs still in civilian service, were much in demand with the increase in ship movements, and were a truer growth area. By Alexandra's standards, as fundamentally a dock and river towing concern, they became long, frequent, and lucrative: Cherbourg to Sunderland, or a four-tug tow in the dangerous waters of the English Channel from Dunkirk to Le Havre for £2650. Shorter sea tows, as from the Mersey to Glasgow or to Cardiff, were dearer than they had been, because of war risks and escalating costs. The effect on Alexandra's sea towing revenue was startling. In a ten-week period in summer 1915, it averaged over £500. With the boost from sea work, total recorded weekly revenue in 1917 climbed several times to over £3000, or even £4000. In the winter months, when there was generally less open-sea work, revenue could be patchy; but when weekly revenue from the Mersey fleet touched a new record in the first week of January 1918, nearly half of it came from sea towing. Alexandra even contemplated towing a giant, 4200-ton raft of timber from Canada to Britain, but did not, it seems, take the job. It was eventually done by Watkins's *Racia*.

During 1917, shortage of tugs on the other side of the Channel brought the company profitable charters: the *Hornby* went to Rouen for £3700 a month. However, far and away the biggest contributor to Alexandra's revenue in the war years was the Admiralty hiring of the tugs they requisitioned. The rates were not so high as those for private charter, but the frequency and duration of the requisitions earned the company a fortune.

The 'Empress of Britain' and the 'Empress of England' with Alexandra tugs in attendance. Another tug can be seen towing a grain elevator. (F. Leonard Jackson, Heswall).

After the first few weeks the receipts themselves are not recorded with the rest of the revenue, but the figures we have are indicative. By the end of August 1914 commandeered tugs were already bringing in over £1000 a week between them. Hiring seems to have been on a monthly basis at first, with figures of up to £1050 quoted for the big modern tenders *Flying Kestrel* and *Flying Breeze*, £900 for the older *Herald*, down to £650 for the *Wapping*. By 1915 daily rates were spoken of - between £22 10s and £13 for tugs on examining duty. As a consequence, we find a net profit of £4777 for the first half of 1915, the highest recorded figure to date. It was quickly eclipsed: the second half-year netted £5697, a 25s bonus was paid on each share, and the grateful shareholders paid the directors £1000 between them, in addition to the usual £500 fee. The directors did not, however, see as much of their fees as they used to: income tax went up to 3s in the pound in 1916, and to 5s in 1918.

Net profit figures thereafter are not known; but the trend may be glimpsed from other indicators. Revenue from all sources, £36,500 for the first half of 1914, had almost doubled a year later, and more than trebled by the second half of 1917, its wartime peak. Transfers from profits to reserves over four years, including provision for taxation, totalled £192,000. This build-up of reserves would pay dividends - literally - in the lean years to come. Bonuses of 12s 6d per share and yearly and interim dividends of 3s 9d per share became normal. And finally, Alexandra found themselves turning over large sums to the Exchequer by way of Excess Profits Duty, introduced in 1915 as a wartime measure, under which firms were allowed to keep all profits if they were no more than 20 per cent higher than a pre-war average, but forfeited a proportion of any further profit, which by 1918 was 80 per cent. In the first year of hostilities Alexandra paid nearly £17,000, and in 1918 the figure was almost £40,000.

Encouraged by the wartime boom, a new towing concern was set up by the redoubtable James H. Lamey. He had been a pilot on the Manchester Ship Canal, said to navigate his way along it by throwing lumps of coal into the water ahead of his ships: when there was no splash, he changed course. Equally remarkable was James's brother and partner William, another canal pilot. A drinking man, he was alleged to say of himself that even after hitting the bottle, he was able to make an upside-down drawing of a full-rigged sailing ship accurately. One of James's sons, Alfred, was given to what was called 'unorthodox' behaviour, that earned him the present of a pirates' Jolly Roger from other Mersey tug owners. Unabashed, he flew the flag from his masthead.

Having also held a master's certificate for sail, and owned sailing coasters, James Lamey was as well qualified as anyone to assess the potential of towing work on the Mersey. In 1916 he bought the little old 25-ton paddler *Hero*, and set up the Hero Tug Company to operate her. Astutely, he put her to work servicing the anti-mine measures equipment installed on merchant ships in the Mersey. In due time, Lamey's tugs would come under the wing of Alexandra; but for the present, their owners brought a dash of character to the river that was all their own.

Other Merseyside firms than Alexandra were spreading their wings, encouraged by the wartime boom. The West Coast Towage & Salvage Company was set up in 1917 by Prendiville. In three years it mopped up several small concerns, including the Star Tug Company, and the appealingly named Steam Tug Expert Company and Sunshine Tug Company.

8

Bramley Moore 1922

Depression Years

A tide of false prosperity carried shipowners and tugowners into the first postwar years. International trade, starved of transport by the overriding needs of war, generated a vast demand for ships. Those mass-produced during the war to replace losses could not satisfy it for, though plentiful, they were of a standardized, basic design useless for specialized peacetime trades, in which the layout of a cargo liner depended on the nature of its freight. Because ships were scarce, out-of-date rust-buckets, cherished in wartime, were kept in service still longer. Anything that would float made money. Shipbuilding boomed, in spite of the high price of labour and materials. A 7000-ton tramp built for £42,500 in 1914 now cost £169,000. Cargo liners cost two or three times as much to build as tramps.

The result was an armada of ships that were old or otherwise unsuited to postwar conditions, or else new and very far from paying for themselves. Their numbers were swelled by construction in countries that had never had large merchant navies of their own - Italy, Japan, the United States - and which had felt the lack during the war, when the foreign ships on which their trade depended disappeared. Japanese construction went up by 50 per cent in the war years, and that of the United States by no less than 500 per cent. After the war, while ships were still in short supply, freight rates stayed high, and the governments of these countries came to believe that a big merchant fleet was necessary, not just for security, but also, like the modern airline, for prestige. Not being the product of economic laws, these fleets had to be more or less state-subsidized, and therefore presented other shipowning nations with loaded competition. The Japanese, furthermore, paid their workers two-thirds less than their counterparts in the West, making their ships cheaper to build. Later, Germany and France joined the nations with subsidized, semi-nationalized shipping industries.

In another manifestation of 'economic nationalism', countries that had been cut off from imports during the war - India, Australia, Canada, Japan

'Belfast' approaching the Thames Barrier on her way back to her permanent mooring in the Pool of London by Tower Bridge. (Handford Photography, Croydon).

'Crosby' with Atlantic Container Line's new 37,200 dwt third generation ship 'Atlantic Companion', June 1984. (Hal Mullin, Bromborough).

The Tall Ships on the Thames with 'Sun Anglia' in attendance (J. Godsell, Wallasey).

A bevy of tugs around the 'Burmah Endeavour' in Southampton.

'Formidable' and 'Indomitable' built in 1979 with a bollard pull of 55 tonnes, were intended to handle the biggest ships afloat, construction barges and offshore drilling jackets. (Fotoflite, Ashford).

'Brightwell' with stern-drive, twin Aquamaster propulsion assisting the 29,411 grt 'Gulf Spirit' with 'Alfred' in Felixstowe.

'Invincible' is just visible as she services Total 14/9 MCP-01 in the North Sea. No place for faint hearts!

HMCS 'Margaree', a visitor to Liverpool, with Alexandra tugs in attendance. (J. Godsell, Wallasey).

'American Pioneer' leaving Felixstowe with 'Ganges' ahead and 'Sauria' on the starboard bow. 'Alfred' and 'Egerton' in the foreground.

'Bramley Moore' attending H.M.Y. 'Britannia' on the occasion of the Queen's visit to Liverpool in August, 1988. (B.S. Clarke).

Voith Schneider tractor tugs in action with an ACL container ship at Gladstone Lock. (Elsams Ltd, Liverpool).

'Bramley Moore' again, this time assisting the British Gas Accommodation Rig AV1 which was built at Cammell Laird, Birkenhead in November 1984 for the Morecambe Bay Gas Field. (Hall Mullin, Bromborough).

The twin-screw Schottel tractor 'Flying Kestrel' assisting the 'Queen Elizabeth 2' with 'Albert' in Southampton Docks.

'Collingwood' and 'Canada' assisting a submarine in Gladstone Lock. (P.B. Parker, Formby).

3444 b.h.p. pushing 'Bramley Moore' through the water. Built by McTay Marine Ltd., 1984, she is engined by Ruston Diesels Ltd., 38.5 tonnes bollard pull. (Hal Mullin, Bromborough).

Built in 1987 by Richards (Shipbuilders) Ltd., 'Brightwell' named after a village near Felixstowe, has twin Ruston Diesels driving Aquamaster units. (Brian Ollington, Great Yarmouth).

The first rig for the Morecambe Bay Gas Field, Western Apollo II, with 'Wellington' and 'Wallasey' in attendance.

'Redoubtable' bought in 1982 from France, anchor handling tug with a 70-ton bollard pull and 5000 hp.

- started their own industries, and so were no longer dependent on those of traditional exporters such as Britain. They erected tariff walls to protect their new industries when imports resumed, further discouraging trade. There was in fact little more international trade than there had been before the war, but a great many more ships - far too many for the available business. For the tramp steamer, a fall in Britain's coal exports - from 173 million tons in 1913 to 40 million tons in 1937 - was a particularly bad omen.

In Britain, the decline in shipping began in 1920, when wartime controls on freight rates, which had kept them artificially high, were swept away. The bottom dropped out of the freight market, shrivelling the merchant marines of countries without subsidized fleets. The same ton of coal that had cost £10 to carry to Egypt in 1918, cost 11s 3d in 1923. The tramps that customarily carried such bulk cargoes suffered especially, for they operated on reserves that were slim compared to those of the cargo liners. Also, the cargo liner operators, themselves desperate for business, began carrying bulk commodities cheaply, cutting into the tramps' market.

With no incentive to build, British merchant shipping grew older and less efficient, and its tonnage remained at around its prewar level until the 1930s, while the tonnage of other nations' fleets grew by 75 per cent, all of it modern and efficient. In the worst years, between 1929 and 1932, nearly a fifth of Britain's merchant shipping tonnage was laid up. Passenger liners were stacked three abreast in the shelter of the River Dart. Dividends slumped: in the boom year of 1912, when British ships were carrying half the world's seaborne trade, a cross-section of shipping firms had paid an average of 7 per cent. Another sample, for the years 1921-34, gave an average of 4⅔ per cent. For ships' crews, the effects of the slump were even more drastic. One ship sailed from Bristol with six holders of masters' certificates serving as able seamen; and they were the lucky ones. Their colleagues washed up in restaurants, or joined the five hundred other masters who applied for a single harbour master's job. Engineers were better off: their manual skills enabled them to get jobs turning coffin-handles.

As ever, the fortunes of British tug operators mirrored those of their customers. In the aftermath of war there was at first too much work for the available tugs. The Admiralty released its requisitioned boats slowly, and their owners then waited until the Admiralty had reconditioned them at its own expense. Tugs left in civilian service were worn out, for there had been little time to keep them in repair. Then, no sooner had operators got all their tugs back, refitted them, and built new ones at great expense, than the shipping industry went into decline. The drop in business was exacerbated by labour troubles arising from general economic slump and by excessive competition.

The effect of labour disputes could be shrugged off by all concerned in times of plenty; not so when times were hard and margins narrow. Decline had already set in on the Tees when Crosthwaite's two hundred tugmen came out for four months between October 1920 and January 1921. They wanted more money, and claimed that their conditions were 'the worst in the country'. Crosthwaite counter-claimed that their pay was better than anywhere else in the north-east, and that, in any case, engineers were just 'drivers', who deserved no more. The tugmen lost, and only fifty got their jobs back, for ships had gone elsewhere and the business was lost. The General Strike of May 1926 brought the Tees tugmen out for nearly a month, but at the end of it, there was only enough coal to run four Crosthwaite tugs out of eighteen, owing to the miners' strike that went on for seven months, so again, the tugmen lost.

On the Thames, there was a lightermen's strike in 1923 that hit barge operators such as W.H.J. Alexander, Charles Gaselee, and J.P. Knight; while a renewed rate 'war' between lighter owners again forced towage rates down and cut tug revenue. As on the Tees, the General Strike was dominated for tug owners by the shortage of coal it entailed. A few Watkins masters and most of their engineers stopped work, but crews could be found by enlisting office staff - even the head of the firm, John Rogers Watkins II, who skippered a tug. But when a tug had to sail to a continental port to coal for work on the Thames, the cost bore no relationship to sound economics.

The towing industry had never been so competitive and fragmented as in the years up to the early 1920s, with around 100 businesses employing a total of about 8000 crewmen, but the new economic climate could not sustain unrestrained competition and fragmentation. The first impulse for a change appeared on the rivers of the north-east. Competition was many-headed on the Humber, where in 1914 Hull Associated Tug Owners was founded by five operators pooling forty-two tugs. At the end of 1920 several firms, with thirty-nine boats between them, amalgamated to form the United Towing Company. In the same year William Crosthwaite on the Tees, then a declining port, put together the Tees Towing Company from his own Tees Tug Company and the Robinson company, creating a fleet of twenty boats. His takeover policy brought Crosthwaite accusations from the shipowners of monopolistic ambitions, price fixing, and high towage charges, particularly in 1929, when he absorbed the last remaining commercial operator on Teesside. Also in 1920, although on a much smaller scale, there was amalgamation on the Tyne as well, when John Batey & Son linked up with the Lawson Steam Tug Company to form Lawson - Batey Tugs Ltd. All these companies survived through bitterly hard times, a testimony to the virtues of rationalization that was not lost upon the Alexandra Towing Company.

By the early 1930s the shipping slump was hitting tug operators hard. The drop in business simultaneously cut their revenues and intensified the competition between them. As in shipping as a whole, too many boats were chasing too little trade. Rationalization by amalgamation was one approach to the problem; another - a more ambitious version of an existing idea - was co-operation, embodied in the formation in 1934, at the instance of William Crosthwaite, of the British Tug Owners' Institute (soon to be renamed the British Tug Owners' Association).

The Association's aim was to discuss, promote, and concert action in the common interest of tugowners; in fact it was a port tug association writ large. Crosthwaite sought a set of standard towing conditions, applicable throughout the industry, to cut down mutually destructive competition. But the only standardized tariffs were those instituted within a port, and even they tended to be ineffective. Many tug proprietors disliked united action on principle, in case it limited the individual owner's freedom of action. The Association left labour relations strictly to local owners to handle.

A London Tugowners' Association was formed in 1937, but it had little to do with standardizing rates. It arose from pressure on the tugowners by Thames shipowners to resume relations with the Transport & General Workers' Union, suspended after the General Strike. The tugowners acquiesced - indeed, by December 1936 the United Steam Tug Company was advising its masters to join the union - but they created the Association as a measure of self-defence against both shipowners and the union. It was also to evolve a standard tariff, but this was incidental.

By this time the state of shipping, and therefore the prospects for tugs, was improving. The shipping industry was fighting back by means of drastic slimming measures: five million tons of old ships were scrapped between 1931 and 1937, and not replaced, while in the latter year the number of seamen it employed had fallen to 130,000, from a 1912 figure of 210,000. Two trades were actually expanding: the holiday cruise provided some passenger liners with a living - indeed, liners such as the *Arandora Star* were built specially for this trade - and the number of oil tankers grew worldwide from 258 in 1912 to 1558 in 1937, by which time they aggregated 10 million tons. They made up 15 per cent of the world's merchant tonnage, and nearly 10 per cent of Britain's.

An upturn in trade began in 1937. That year, it was 30 per cent higher than in 1914; though the tonnage of shipping available to carry it was still 40 per cent greater. By this time Britain's share of the world's merchant shipping tonnage had fallen from 41 per cent in 1914 to 26 per cent. The British merchant navy was 1½ million tons smaller than in 1914, while all other fleets except that of Germany had grown.

Generalizations about the towing industry between the wars can go only so far: different ports had very different experiences. The fates of London, Teesside, and Liverpool, for example, contrasted sharply.

Like those elsewhere, London tug operators found business booming when peace returned in 1918. Ship tows multiplied; and for those firms such as Gaselee who towed lighters as well, turn-round was faster. Unlike operators elsewhere, however, they did not then fall into bottomless depression. The decline in shipping was not so pronounced on the Thames as in the North, and revival in the 1930s began sooner because the Thames won most of the benefit from the few trades that were doing well. In 1909, 38½ million tons of goods valued at £322 million had passed in and out of the port of London. In 1937, the figure was 62½ million tons, worth £605 million.

There was a local boom in traffic serving the growing demands of the motor vehicle, concentrated on developments lower down the river. In 1924 a new oil refinery was opened on the Medway, bringing more work for Medway and Thames tugs. To the existing oil terminals at Thames Haven below Gravesend, a refinery was added in 1921. At Shell Haven next door, another refinery had been built in 1916. At Coryton, a refinery was opened in 1921, and an oil storage terminal arose on adjoining Canvey Island in 1938. Further upstream, oil wharves were built at Purfleet for British Petroleum and the Anglo American Oil Company. More tankers - already awkward to handle because of their length and the fact that most had only a single screw - needed more tugs. Transhipment of oil and petrol upstream to London was done by lighter and tug. And in 1933 the Ford Motor Company opened its vast new factory at Dagenham, bringing iron ore bulk carriers to its wharfs. New building on this scale was foreign to other ports. One of its by-products was to complete a shift of the centre of tug operation downstream from London to Gravesend. The movement had begun with the seaward spread of the docks, and left only the lighter tugs still based upstream.

Another factor in the relative prosperity of the Thames was the big passenger and cruise liner. Most docked at Tilbury, but others needed tugs to help them further upstream, to the Royal Albert and Royal Victoria Docks, and to the new King George V Dock opened in 1921. The ships of the Union Castle line came to East India Docks, and Cunard, Canadian Pacific, and the transatlantic vessels of Furness Withy as far up as the Surrey Commercial Docks. Passengers liked to disembark as close as possible to the heart of the capital.

Trade on this scale justified new building programmes such as that initiated by Gaselee in 1921. Six tugs were launched in ten years, all over 100 tons and of up to 550 hp. Three tugs were bought and five were broken up, leaving the company in 1932 with a modern fleet of fourteen boats, more powerful than those of 1921. Gaselee became pioneers of the diesel ship tug on the Thames when the new 82-ton *Agama* started ship towing in 1938. By 1939 the fleet consisted of six diesel tugs, mainly for barge work, and ten steam tugs. The fleet of Gaselee's principal competitor, J.P. Knight, old and ill-kept, was in poor shape by the early 1920s; but times changed for that firm too. In the 1930s its strength rose from ten to seventeen tugs. Between the wars W.H.J. Alexander moved more strongly into ship towage, now towing to and from the continental ports. Alexander himself died in 1929. His six tugmaster sons and two daughters inherited the business, which the eldest son, George, ran. In 1938 the lighterage side was sold - four tugs and 110 lighters - though Alexander remained wharfingers. Seventeen river and sea-going tugs constituted the fleet in 1939. The Gamecock Steam Towing Company was taken over by the Ocean Salvage (later Salvage & Towage) Company in 1928, but retained their identity as dock and river tug operators. A fleet of five tugs, most of them modern, bore the blue and black funnels in the tideway in 1940: the *Water Cock, Crested Cock, Ocean Cock, Atlantic Cock,* and *New Storm Cock.*

Watkins, too, benefited from the flourishing river work on the Thames. In 1923 their fleet numbered fourteen boats. When the price of building fell, owing to declining demand, they ordered the 200-ton *Gondia* and *Kenia.* Delivered in 1927, they had the power - around 760 hp - to deal with liners and tankers; and they carried passenger certificates enabling them to tender. The *Kenia* was fitted with wireless in 1929 - the first Thames tug to be so equipped after coming into commercial service. In 1931 another very modern, general-purpose boat, the 850 hp *Tanga,* was added. The United Steam Tug Company of Gravesend - 'Ring Tugs', after its funnel insignia - was owned by pilots, who ran four tugs. Two - the *Dendera* and *Dongora* - were new, built in 1930 and 1932 respectively. In 1937 all four were taken over by Watkins, bringing their strength up to twenty-two boats. The *Dendera* and *Dongora* were renamed *Racia* (the first of the name having been long since sold) and *Muria.*

Optimistically, Watkins built their big new river tugs strongly enough for ocean salvage work, and gave them salvage pumps; but in this area Watkins's day was done. They had tried a come-back just after the war when they acquired a 286-ton, ex-Admiralty, ocean-going tug called the *Dreadful,* which they sensibly renamed *Rumania.* She had triple expansion engines developing 1000 hp, and retained her wireless apparatus. But Watkins could not find enough work for her, and she was sold in 1923. Nowadays Watkins tugs seldom sailed further afield than the Downs in the south or the Sunk light in the north. Dutch and later German operators reigned supreme in ocean towing, the only realistic British opposition coming from the United Towing Company of Hull, who worked out into the Atlantic and Mediterranean. In 1925 their purpose-built deep-sea tug *Seaman* towed two ships to Buenos Aires, earning £3400 in 124 days at sea. From 1928, Crosthwaite's Tees Towing Company, in search of new business, diversified into sea towing with a Dutch tug they renamed *King's Cross;* but they competed much nearer home, in the North Sea and Channel. Four years later they added a boat originally built for the Dutch, the *Queen's Cross.*

Throughout most of the inter-war years, the Tees furnished a picture of economic disaster at the opposite pole from the Thames. In 1913 the river had cleared 5644 ships in and out - a figure that fluctuated thereafter, but tended remorselessly downwards. In 1919 it was 2100, climbing to 3513 in 1923. But the total was under 3000 again in 1930. Seven years later trade was at last looking up, with 4222 ship movements; but the total was still well down on prewar days.

Crosthwaite sailed into the peace with fourteen tugs. By 1923 the figure was twenty-three, which was far too many for the trade. Wholesale scrapping of old boats followed. Including tugs taken over, there were only twelve on the strength in 1930. Of the eight sailing under the Tees Towing Company name two years later, half were laid up; three years after that, there were only eight boats in the company's combined fleet - a drop of nearly two-thirds in a dozen years. There was nowhere to go on the Tees but up.

Liverpool never again saw boom on the pre-1914 scale, but while it did not do nearly so well as London between the wars, its troughs were shallower and shorter than those of the north-east. Owing to foreign competition, the most important of the exporting industries that were Liverpool's livelihood was in full decline. Cotton goods made up 55 per cent of the city's exports in 1922-4, but only 36 per cent a dozen years later. It was fortunate that the Mersey's share of other exports went up in partial compensation - chemicals, and manufactured iron and steel articles. So, too, did Merseyside's re-export trade. Overall, the value of Liverpool's exports declined from £480 million in 1920 to £103 million in 1932 even while the city remained Britain's foremost export outlet. In the early 1920s Liverpool's import trade suffered a collapse. The Mersey handled 36 per cent of Britain's imports in 1918, but only 22 per cent four years later. Over the years, Liverpool lost import trade by around 1 per cent per year. The gainer from Liverpool's loss was London, which offered the Midlands manufacturers lower transport costs. Merseyside's share of the whole trade of the nation declined from 31 per cent to 21 per cent in 1938 - by which time London had seized 38 per cent. For a city that made almost nothing and existed by its trade, this was the worst of news.

Mersey shipping reflected the decline in the river's trade. In 1920 the Mersey handled 16.5 million tons of shipping, or 17,100 vessels. This was a slight improvement on 1914, and the peak of a short-lived recovery that happened in Liverpool as elsewhere. After the slump that followed, traffic grew once more to 21.3 million tons and 20,800 ships; then in the full-scale collapse of the early 1930s, it fell again to 18.75 million tons and 17,000 ships in 1933. The figures did not recover to the 1930 level until 1937 - and even then, were not two-thirds of those of 1913. It comes as no surprise that the

Gladstone Dock was not completed until 1927. By that date, Merseyside had 658 acres of wet dock and thirty-eight miles of quays, a total that would never be exceeded.

Life was correspondingly bumpy for the Liverpool tug operators. When the postwar recovery ended, a famous old name, John Prendiville & Company, was left high and dry. Their West Coast Towing & Salvage Company vanished from the scene in 1924. By this time the Liverpool Screw Towing & Lighterage Company was down to nine 'Cock' boats. They did all the launch towing for the Cammell Laird shipyard, which built their tugs, but suffered from the takeover of their other old standby, the Allan line, by the Canadian Pacific Railway, who gave the Allan towing to their own Mersey Towing Company. But they benefited from an increase in canal traffic to Manchester, in which they (unlike Alexandra) were still involved; and in 1929 they built their first new tug for twenty years, the *Thistle Cock*.

Newcomers were more active. James Lamey had set up J.H. Lamey & Company in 1920 and expanded in a modest and typically eccentric way. He added two elderly tugs to his existing pair; the *Resolute*, built in 1891, and the *Iris*, a year younger and originally constructed as a millionaire's yacht. The Rea Towing Company's 'Garth' fleet launched a determined attempt to carve out a slice of the Mersey towing business for itself. Of their ten tugs in 1923, six were new - a clear sign of their commitment, although they at first presented little threat to Alexandra, now the biggest tugowners on the river with their thirty-four-strong fleet.

Up to the end of 1920, business for Alexandra's tugs was good, as might be expected. In January 1919 weekly revenue averaged nearly £3700; a year later the figure was over £5000. Income was boosted further by Admiralty hiring fees, for Alexandra's commandeered tugs were not all derequisitioned until early 1920.

There was frequent and useful sea towing - as of a floating dock from Scapa Flow to Liverpool for £2390, and the chartering of the *Sarah Jolliffe* by the London Salvage Association for £120 a day. Sea towing continued to bring in £400 to £600 a week even in the bad years that were coming. Salvage was lucrative, for a change. Early in 1919 the *Alfred* and *Morpeth* shared £1000, of which £300 went to the crews, and the *Morpeth* got £1300 to herself in June 1920, of which £450 went to her crew.

Business was good enough, throughout 1919, to allow substantial transfers from profits to reserves in spite of the institution, in April, of a 48 hour working week for Liverpool dockers and their fellow union members the dock and river tugmen. It made no practical difference to the hours the tugmen worked - when necessary, he still had to work all the hours God gave - but it could make a vast difference to his income, for if he worked for a firm such as Alexandra, he and his 600-odd colleagues were paid at overtime rates for additional hours worked.

In practice, an Alexandra tugman's basic pay in early 1920 covered not forty-eight hours but fifty-four, made up of six shifts of nine hours in a week, in each of which one hour was for a meal break. He was paid overtime above the fifty-four hours. Wartime rises and the new overtime payments between them gave an Alexandra master on the Mersey average weekly earnings, without commission, of £12 18s 6d, a deckhand qualified as able seaman £6 3s 6d, and a boy £3 6s - four to six times more than their 1914 income, without taking salvage into account. In June 1920 there was a further rise in basic pay, rather than the postwar cuts promised. The purchasing power of the pound had halved in the past thirty years; but after reaching a high point in 1920, the price of the commodities the tugman bought fell year by year. If he worked on the Mersey, times were good, and his employer paid overtime, the tugman of the 1920s was much better off than his father on the boats. To pay their increased wage bill, the Mersey tugowners doubled their towage rates in September 1919. They could never have taken such a radical step in bad times. Even more remarkably, White Star, Leyland, and perhaps others agreed to pay the new charges.

Several changes were taking place on the board of Alexandra. John Hughes died in August 1919, and five months later two other directors were elected: Isaac Mack, son of the chairman, and Thomas Barrett Bicket, known as Barry - another of Alexander's sons. At about the same time Silvester Procter was given the job of Company Secretary in addition to that of Manager, which he had held since 1900. From the war years, his name figures more and more prominently in the Alexandra annals as an increasingly valuable, and valued, officer of the company.

Tugs were sold for a good profit - the *Salthouse* to Elder Dempster, the *Herald* to the Anglo American Oil Company for service in the Persian Gulf, the *Sloyne* to Watkins - but there were a lot to bring back into service, and only two newcomers joined the fleet, both second-hand: the *Neptune* and *Hodgson*. This was fortunate, for by September 1920 four tugs were laid up for lack of work. By November the number had grown to five, a drop in revenue had been noted, and the company was asking its crews to accept a wage cut, only five months after the last rise. By May 1921 a dozen Liverpool tugs were out of service, their crews laid off. Over the next five years there were never fewer than three Alexandra tugs laid up on the Mersey.

Difficulties multiplied, and compounded one another. Having fallen back after the war, the price of coal shot up again - reaching 36s a ton in mid-1919, and 45s in November 1920. The fleet's coal bill for July 1921 was £5000, having averaged only £3600 a month for the previous six months. After further fluctuations the price fell back to about 17s a ton in 1925, but never dropped to its prewar levels again. With less business, towage rates had to come down, accelerating the decline in revenue.

Growing competition for the dwindling custom helped to force rates down. Prendiville were a passing worry in 1921; more ominous was the fact that the Rea Towing Company took delivery of six new tugs in 1923. A new force had arrived on the Mersey. Both Rea brothers had died during the war, and their varied interests had been taken over in 1920 by William Cory & Son, Thames-based coal shippers and collier and lighter owners who ran small tugs on the Thames, and in 1921 had built the Coryton oil refinery on the Essex shore. With this powerful backing a new Rea Towing Company began to seek Mersey towage much more aggressively. Most of their tugs were new, and the most modern in the port. In the eyes of shipowners, Alexandra's big but mostly ageing and well-worn fleet fared badly in comparison.

A crippling rate war was the outcome. By early 1923, Rea were offering to dock ships for £8 a tug, which undercut Alexandra by £3 or more. When it looked as if Alexandra would lose the business of Henderson, Lamport & Holt, MacIver, Houlder Bros, Ellerman, and the PSNC (Pacific Steam Navigation Company) - as was the case by spring 1924 - they had to cut their rates to stay in the race. A reduction was better than losing trade, perhaps permanently; even when Rea's rate went down again to £6 10s, Alexandra hung on, thanks in part to old friends such as Booth, who were willing to pay a little more for the Alexandra name. They did not always have to match the Rea rate; they could keep up even while asking more for the same work, so long as the difference was modest. At the same time, Alexandra and Cock Tugs approached Rea with a view to making a work-sharing arrangement that would remove competition. This Rea rejected at first, countering with a proposal - rejected in turn - that would leave Alexandra with just under half the dock towage business. Psychological warfare played a part. In March 1925, Rea offered to buy Alexandra for £120,000, or £12 a share. Alexandra replied with a counter-offer for Rea, which was no more serious.

Only two new boats were added between 1917 and 1928, and in special circumstances. One was another *Bramley Moore,* delivered by Cran in summer 1922. This £16,000 order may have had less to do with demand for towing than with the condition of Cran, who, together with the rest of the shipbuilding industry, were suffering from lack of work. With their 25 per cent interest in the firm, Alexandra had an interest in keeping Cran solvent, and had already guaranteed a £20,000 overdraft for them. When Alexandra ordered another tug from Cran late in 1925, there was no doubt about it. Sir Alexander Bicket, knighted for his wartime services, lent Alexandra two-thirds of the cost out of his own pocket, specifically to help the foundering Cran with an order. The latest *Wellington* was a big, modern tug of 285 tons, 113 ft long, with triple expansion engines delivering 1000 hp and boilers working at 200 psi. Generally, however, tugs were being put up for sale

- the older ones, at least - in an attempt to match the size of the fleet to that of the shrinking market. The last Jolliffe tug - the *Sarah Jolliffe* - was sold in 1925, and the company wound up. Another famous name had vanished from the Mersey.

Alexandra's difficulties showed in the incomes of the business, the shareholders, and the workforce. The figures told a story of remorselessly growing expenditure and shrinking revenue. Results for the first half of 1921 were described as 'deplorable': revenue was 30 per cent down on the corresponding period of 1920. By late 1921, the income from all sources was down in most weeks to between £2500 and £3000, while expenditure on wages and coal had diminished by less than half as much. Thanks to high operating costs, income that would have been thought miraculous before the war was now disastrous. The third quarter of 1921 saw a loss, and for the first time in the company's history there was no dividend in that half-year.

The decline accelerated into 1923, profit margins hovering between the paper-thin and the non-existent. A modest net profit for the third quarter of the previous year was turned into a loss as the Rea war hotted up. The second half of the year showed a tiny profit, and the see-saw of alternate profit and loss continued into 1924 and 1925. In the worst spells the directors forewent their fees; although almost always, however bad the results, there was a distribution to shareholders out of reserves. The company's reserves and investments, built up during the war, were now its salvation. It was able to run up an overdraft at the bank; on one occasion it made up a half-year's loss out of reserves: and on others, it had War Loan stocks to sell to enable it to maintain solvency.

Bad times for the owners and shareholders inevitably meant bad times for their crews. There was confrontation in February 1922, when the Liverpool owners told their crews that they would have to work longer basic hours (already up to fifty-six), so earning less overtime, and take a cut in basic pay and commission. The crews came out on strike for nearly two weeks; but they were on weak ground, for out of sixty-odd Mersey tugs at work in good times, half were laid up for lack of work, and the owners kept all but five of the rest operating with outside labour. The tugmen's union - now the Transport & General Workers' Union, which had absorbed the dockers' unions in 1921-2 - could do nothing: the pay of Liverpool crews was cut by between 14 and 25 per cent, the ordinary seamen and trimmer coming off worst. Officers' commission was cut back to its prewar level of 2½ per cent. Alexandra had managed to keep five tugs manned during the strike, but revenue suffered, falling by nearly three-quarters during the first week.

There was worst to come. Most crewmen had to accept three more cuts in pay in 1922-3, although strikes won back some of it. The men came out three times, too, against the laying-up of tugs and laying-off of crews, with some success - tugs were put back in commission. Rea, as always, acted independently: their crews worked when Alexandra's did not; and when increases were given, Rea paid them sooner. Some idea of the significance of the wage element in Alexandra's expenditure is afforded by the figures for one week in September 1924. Out of total outgoings of £2775, wages account for £1477 - over half. Coal is the next biggest item, swallowing up £482. Then come repairs, taking £317 in that week - the penalty of running a large fleet of old tugs.

The most spectacular strike of all ended in total defeat for the tugmen. When the Transport & General Workers' Union called them out for the General Strike in May 1926, Alexandra kept three Liverpool tugs in service with non-union crews and volunteers; a fourth ferried strike-breakers to and from ships in the port. Revenue fell overall by about half in the fifteen days of the strike. While the miners stayed out the price of coal rocketed - 34s a ton by June, up to 70s a ton in October - and it was coal of poor steam-raising quality, at that. But then the miners' strike ended too, and the bottom dropped out of the price of coal: by November it was 19s a ton. The tugmen had gone back on 18 May with nothing won either. Militancy among the crews was exhausted for years to come: a fact of which the tugowners took advantage by refusing, henceforth, to recognize the union. Alexandra demanded, and got, agreement from their masters and engineers that they would no longer be members of the union.

Meanwhile, the company had established itself in two other bases - Southampton and South Wales - to some degree off-setting the tribulations they had been experiencing on the Mersey.

Wallasey 1954

9

Outports and Recovery

Southampton had been Britain's foremost port as long ago as the Middle Ages, thanks to unique geographical blessings. Southampton Water offered ships a straight, deep channel eleven miles long and up to two miles wide, sheltered by the Isle of Wight from the seas and gales of the Channel. The range of water levels between high and low tide was comparatively small, making locks at dock entrances unnecessary. Most remarkably, the position of the Isle of Wight gave Southampton four high and four low tides a day instead of the two with which Britain's other ports had to make do. This meant that there was enough deep water at the berths for ships to tie up at any time of the day or night. The high tides came in pairs, only two hours apart, each giving two hours of high water.

In spite of these advantages, Southampton lost trade to London from the sixteenth century, but - in common with so many ports - began to come to life again at the end of the eighteenth. Like London, it was an entrepot - a funnel through which goods and passengers were distributed nationwide, with almost no industry or other *raison d'etre* of its own, and no immediate industrial hinterland to feed it with exports or absorb its imports. Unlike London, however, Southampton was not big enough to provide its own market for goods; it was totally dependent on its traffic.

A harbour board was formed to improve the port in 1803, but nothing much more happened until links with the rest of the country improved with the coming of the railway in 1840. Southampton then became a point of embarkation and arrival for passengers using some of the new steamer routes to America and to the East. The Royal Mail Steam Packet Company moved in in 1839, and the P&O line a year later. Their big steamers had at first to moor in Southampton Water and receive or discharge their passengers by small craft. Docks were called for, and in 1842 the 16-acre Open, or Outer, Dock was opened. The Close, or Inner, Dock, with 10 acres, followed in 1851.

Major expansion began in 1890, when the London & South Western

Alexandra bought the massive 2600 h.p. Admiralty tug 'Rollcall' as a passenger tender. She was renamed 'Romsey' in 1922 after being converted to oil firing. Sold to French Line in 1929.

'Harrington' built by J.T. Eltringham in 1903. She sank while towing a tanker in the King's Dock at Swansea in 1954 but was raised and broken up in 1959 (Albert Hayes, Salford).

Railway took a hand in the building of the new Empress Dock, of 18½ acres. By this time the dock company was in a bad way, and two years later it was taken over by the railway company. From that moment, Southampton never looked back. The P&O had returned to London in 1881; but by 1893 had been replaced by the transatlantic steamers of the American Line (formerly Inman), and of the North German Lloyd and Hamburg-Amerika lines, which offered a route from the Continent and southern England to North America independent of Liverpool. The L & SWR made a determined attempt to win more of the expanding transatlantic passenger trade away from Liverpool, tempting shipping lines not only with the peculiar advantages of Southampton, but with a reminder that the fuel of the South Wales coalfield was only 129 miles away by rail via the Severn Tunnel. They succeeded dramatically. In the single decade 1890-1900, the total tonnage of ships in and out of the port increased by two-thirds. And it was not only passengers that the railway company was attracting: in the following thirteen years, the value of goods imported and exported through Southampton rose by £30 million a year.

The town won its first great coup in 1907, when White Star transferred four of its crack transatlantic liners from Liverpool in order to compete with the German liners working out of Hamburg. By 1911, Cunard ships plying between London and Canada were calling at Southampton, too, enabling passengers in a hurry to reach London in a couple of hours. Dock development had to follow the newcomers quickly: the opening of the 15½-acre Ocean Dock in the same year of 1911 gave even the biggest ships enough water under their keels - 40 ft at low tide. The Itchen Quays, opened in 1876, were extended in 1895 and the Test Quays and the South Quay had been added to the berthing accommodation by 1902. By 1913 the L & SWR had spent £6 million on the port, which was now the sixth busiest in Britain. Such a rate of growth was unmatched by any other port - all others, Liverpool included, gave the impression of relative decline when measured against their new rival.

Finally, with the end of the First World War, Cunard followed White Star south. Although they retained a Liverpool presence, the cream of their fleet, in terms of speed and reliability, moved to Southampton, including their transatlantic mail service. The move was part of a shift away from a Liverpool-centred, passenger-centred orientation that had begun before the war. With the move of the two big transatlantic passenger line operators, passengers could board or land at a point much closer to London - a mere seventy-eight miles away. Furthermore, passengers to and from the Continent could board and land conveniently at Cherbourg or Le Havre,

'Alexandra' came into service in March 1917, and was requisitioned for war service in December the same year. She continued to give good service until broken up in 1962.

where the liners called on their way to and from Southampton. These were particularly strong selling points with Americans, most of whom made for London or mainland Europe. There was no such bonus for operators on the Liverpool route. Liners could call only at Irish ports, where there were few free-spending first-class passengers waiting to board.

As elsewhere, steam towing had followed close behind the steamship. By 1820 the 17-ton steamer *Earl of Egremont* was towing barges for a local canal company between Milton near Portsmouth and Dell Quay on Chichester harbour. Five years later steamboat operators were offering to tow sailing ships in and out of Ryde, across the Solent on the Isle of Wight; and in 1828, steamers were towing lighters laden with passengers, vehicles, and animals between Southampton and the island.

The first purpose-built tug at Southampton was the *John Lee* of 1844, built for the Southampton Dock Company, which subsequently did service with Watkins on the Thames. And the first vessels into the new Inner Dock in 1851 were the new screw tug *Mary*, built for Andrew Lamb, Superintendent Engineer of the P&O, followed by the P&O liner, *Madrid*. Regular towage began in 1853, with the formation of the Southampton Steam Towing Company. Their first tugs were the 83-ton *Aid*, built in 1852; the 71-ton *Phoenix* of 1853; the *Monarch*, 122 tons and iron-hulled, which joined the fleet in 1854; and the *Belmont* - she of the eccentric adventure in Southampton Water. The company was renamed the Southampton New Steam Towing Company in 1870, and under that name acquired the *Alexandria*, a 110 ft iron screw tug of 121 tons.

'Crosby' attending the R.M.S. 'Mauretania'. She also stood by during the efforts to raise 'Thetis' in Liverpool Bay, 1939.

Nine years before, the ponderously named Southampton, Isle of Wight and South of England Royal Mail Steam Packet Company had been founded, becoming famous for its passenger paddle steamers plying between Southampton, Portsmouth, and Cowes on the Isle of Wight. In the next century its fleet became better and more handily known as the Red Funnel boats. The company's founder was the same Andrew Lamb who had worked for the P&O. In 1884, Lamb's concern diversified into towing by taking over the Southampton New Steam Towing Company - by then working three tugs - and in 1885 they built the tender *Albert Edward* for the growing liner traffic. The company built big to cope with this trade. The steel tug *Hercules* of 1890 was a twin-screw boat of 1200 hp and 234 tons, 135 ft long, and the *Vulcan* three years later was a 288-tonner. The *Hector,* in service in 1903, developed the remarkable power, for its day, of 1600 hp. Such boats were thought necessary for handling ships of the order of North German Lloyd's 14,350-ton *Kaiser Wilhelm der Grosse.*

The brothers R. & J. H. Rea came to Southampton in 1895 with their colliers and coaling business, following the boom and working initially for the American Line. They used tugs to pull their coal barges: they did not compete for ship towage.

Between 1906 and 1909 Alexandra conducted negotiations with the Union Castle, Royal Mail, and American lines, and with the London & South Western Railway that owned the docks, by which tugs might have been sent permanently to Southampton: and not surprisingly, it was this most spectacular of expanding ports that was the location of Alexandra's earliest outport, where their tugs were for the first time fixed outside the Mersey. When in 1919 Cunard transferred their crack liners to Southampton, they invited Alexandra, who were old friends, to accompany them to do their towing. Alexandra had just been looking, not for the first time, at Middlesbrough, with a view to sending three tugs there; but wisely decided against it. Accordingly, in October 1919 a local office was opened, under the management of Frank Stark, and the *Flying Kestrel, Gladstone, Herculaneum, Wallasey,* and *Harrington* moved in. They were joined early in 1920 by the *William Poulsom* and *Alexandra.* Alexandra's decision to go to Southampton was undoubtedly influenced by the steep wages that had been agreed on Merseyside. Even though officers' wages had always been higher in Southampton, they were never paid commission - perhaps because large-scale tug operation had started there after the end of seeking and tows had never depended on a tugmaster's enterprise.

In a time of general slump, Southampton between the wars maintained its reputation as the fastest-expanding major British port. In 1920 the Canadian Pacific line started working here, in 1921 the new United States line arrived, and four years later the P&O returned after an absence of nearly half a century. Even though the Southampton office had its downs as well as ups - White Star rates had to be cut twice - Alexandra business was good enough in summer 1922 for them to buy the massive, 2600 hp ex Admiralty tug *Rollcall* as a passenger tender. Converted to oil firing, she was put into service as the *Romsey*. The original *Gladstone* was bought back from the Russians in 1923, renamed *Brockenhurst*, and sent first to Southampton. Alexandra's initiative proved a wise one: in the worst years on Merseyside, between 1920 and 1924 the Southampton operation earned the company a steady £200-£500 weekly, dropping in the winter months, rising in the spring, and reaching a peak in the late autumn; and it rose to new heights in summer 1924. Towage contracts were made with Canadian Pacific and White Star as well as with Cunard. During the General Strike of 1926 income from Southampton stayed steady. Downturns there, compared with those of some other ports, were negligible. By 1937 it was the nation's third port, in terms of towage handled: 27½ million tons passed through it that year. Two years later it was Britain's chief passenger port, dealing with 47 per cent of the traffic. Thirty-two shipping companies used the dock and quays, from the great Cunard-White Star line to operators of cross-Channel steamers. In 1921 a small refinery was opened at Fawley; three years later an oil storage terminal followed at Hamble. The tankers of Standard Oil and Shell-Mex began to arrive.

Accommodation for ships grew with the demand. The new Western Docks were opened in 1932, in the following year the King George V Dock was completed, and in 1934 a stretch of deep-water quays 1½ miles long and able to take the biggest ships, was finished. These gestures of faith in the midst of dire national depression were justified by the growth in traffic that followed later in the decade.

Within easy reach of London and the Continent, Southampton attracted more glamorous, newsworthy trade than any other port. The biggest and most exciting transatlantic liners of the day, Cunard's *Mauretania* and their 81,000-ton *Queen Mary*, North German Lloyd's *Bremen* and *Europa*, the Italian *Rex*, the French *Normandie* - all called here to pick up or land passengers. The P&O liners sailed for Australia and the Far East, the Union Castle boats for South Africa.

With the great ships came glamorous people. When the *Normandie* anchored off Ryde to land passengers in September 1936, Gracie Fields, accompanied by the ship's band, sang her famous song 'Sally' from the deck. In the same year, the heavyweight boxing champion Tommy Farr and his friend the ex-boxer actor Victor McLaglen got drunk on board an Alexandra tender and started to trade punches. On another occasion, Charles Lindbergh was smuggled ashore from a tender: the American hero of the first non-stop transatlantic flight was publicity-shy, after the kidnapping of his baby son had attracted the attention of the world's press. Gracie Fields had other links with Southampton. When in 1938 the Red Funnel company launched

'Saxonia' alongside in Liverpool together with *'Reina del Pacifico'* and *Alexandra* tugs standing by. Taken in the late 1950s.

No one could mistake the classic lines of the 'Queen Mary' attended by Alexandra tugs.

a new passenger steamer bearing her name, the star performed the ceremony at Thornycroft's shipyard on the Itchen, singing 'Sing As We Go' as she did so.

The Red Funnel tugs shared in Southampton's boom. By 1929 they had eight tugs and one tender - the old *Albert Edward* - in service to Alexandra's five tugs and three tenders. Their modern boats included the 1200 hp *Canute* and the new 1350 hp *Clausentum*. In 1930, Red Funnel introduced the most powerful tender in the port, the new 1500 hp, 675-ton *Calshot*, capable of carrying up to 922 passengers and up to seven of their cars, as well as baggage, mail, and ships' stores. Red Funnel bought the tug *Paladin*, which accompanied Alexandra boats to the Clyde for the *Queen Mary*'s launch in 1934 and her trials in 1936.

Glamour or oil tankers, every ship was grist to the tugowners' mill. In

A photograph of 'Albert' at Briton Ferry dated January 3rd 1935. Two years previously she had her bottom torn out in Canada Dock and went down in seconds taking three men with her.

a prospering port, there was business enough for everyone. Mutually destructive battles of the kind that took place between Alexandra and Rea on the Mersey were unnecessary, and were avoided by a work-sharing agreement concluded between Alexandra and Red Funnel in May 1924: all existing contracts would be retained and the new work would be shared; an agreement which holds good to this day.

The South Wales ports had powerful attractions for the tug operator, because of their proximity to the South Wales coalfields, where industrially significant coal deposits had been found in the mid-eighteenth century. Cardiff boomed as a coal exporting port when the Taff Vale Railway reached it in 1841. The year before, 195,000 tons of cargo had left its wharfs. By 1874 the figure was 2.9 million tons, and by 1907, 9.8 million tons: a fifty-fold growth in sixty-seven years. With imports, the trade of Cardiff had risen to 11.9 million tons.

Docks expanded to meet the demand. Cardiff's first dock - the West Bute Dock, named after the coal-owning Marquesses of Bute, developers of South Wales ports - had been opened in 1839. Four more docks followed before the First World War - the East Bute Dock in 1855, the Roath Basin in 1874, the Roath Dock in 1887, and the Queen Alexandra Dock, the biggest of them all, in 1907: with 52 acres of space and 42 ft of water, it could accommodate the largest ships. Cardiff's total area of wet docks now stood at 163 acres.

Such activity spawned many tug operators. By the end of 1889, Alexandra had an agent and at least one tug - the *Alexandra* - stationed from time to time at Cardiff, at moorings with a Liverpool crew, towing ships in and out of the port. The company's agent, who was supposed to generate such business, was the target of complaints from the board over the lack of it. We know that the *Flying Kestrel* was working at Cardiff in the following year, for there she added another to the company's tally of accidents. In the 1890s, besides 'outsiders' like Alexandra, there were local businesses such as the Cardiff Steam Towing Company, first in the port in 1857 with their *Iron Duke* and *Lady Bute*. Cardiff docks, like those in most of the big South Wales ports, were owned by the Great Western Railway Company, and there were railway company tugs as well. The dock companies naturally favoured the railway tugs; but there was still fierce competition for the business of the port, and Cardiff was for the time being the only port, other than Liverpool, where Alexandra tugs worked. There they remained intermittently on station, without a permanent base. Returns were low, when anybody bothered to record them. Hostility from local tug operators towards interlopers may have been a factor: Alexandra boats were known to have been sabotaged, by means of nuts and bolts in their machinery.

Swansea, on the River Tawe, had the advantage of a sheltered natural harbour, and was the biggest exporter of tinplate and anthracite (the best

An unusual sight. 'Alfred' is towing the five-master 'Carl Vinnen'. In their time five-masters were considered to be 'superships' (John Clarkson).

'Brocklebank' and 'Wapping' blowing black smoke as they tow 'Richelieu'. A sign of power in those days. (John Clarkson).

quality coal). The improvement of Swansea port in response to its trade build-up began as early as 1791, when a harbour trust was set up. It deepened the Tawe, and built piers. The first wet dock - the Salthouse Dock - was opened in 1824. In 1851 a new channel for the Tawe was cut, and a 14-acre wet dock, the North Dock, was built. In the next half-century the South Dock, Prince of Wales Dock, and King's Dock followed, giving Swansea almost as much dock space - 131½ acres - as Cardiff, and six miles of quays. The amount of shipping using the port jumped from around 5500 vessels in 1865, totalling 700,000 tons, and carrying 1.4 million tons of goods, to 5800 much larger ships in 1913, five times bigger on average, with more than five times the amount of cargo.

The harbour trust, rather than the Great Western Railway, owned the docks until 1923, so competition among tug owners should have been freer than in Cardiff. In fact, around the turn of the century local operators excluded competitors so effectively that they were able to impose uniquely oppressive terms on their customers. For example, since their tugs were small, a large ship was obliged to use two, whether she actually needed them or not. The first tug to 'speak' an incoming ship - which would be one of theirs, while they had no competition - had the exclusive right to the handling of that ship in or out of the port during its call, and if a ship arrived in tow of an 'outsider's' tug, the ship would be charged double the usual tariff for all its subsequent movements in the port.

Local owners included Thomas Milward, Rosser Rosser, and the Britannia Steam Towing Company. The Britannia company, which began work in 1895, would in due time be taken over by Alexandra, after the Liverpool firm had established itself in Swansea. Its little tugs not only did harbour work. The original *Britannia,* which gave the company its name, though only 63 tons and 80 ft long, went seeking too - or 'dodging', as the locals had it. In 1896 the *Britannia* did not dodge quite cleverly enough. She was caught seeking without lights to avoid the attentions of other tugs, and her owners were prosecuted. Later, the West Coast Towing & Salvage Company worked out of Swansea too.

Another business was more typical of the very small tug operations that have abounded in British ports. The Dewsburys were a Swansea seafaring family who owned tugs at least as early as the 1890s, when the boy George Dewsbury was working with his father in the *Contest*. In 1902 his father bought the little river tug *Fawn,* of which George eventually became skipper; and ten years later the *Victory* was on the strength as well. The tugs were largely crewed as well as owned by the family. Mrs S. J. Dewsbury, who was the proprietor by 1912, employed her sons John and Thomas as a master and engineer. This business, too, was ultimately absorbed by the Alexandra Towing Company.

In 1920 Swansea's huge, 150-acre Queen's Dock was opened, as was the

An undignified position for 'South Cock' after a freak accident threw her on to a slipway. This photograph gives a very clear view of the fixed Kort nozzle and the shape of the hull.

nearby Llandarcy oil refinery, to which oil from the British Tanker Company's ship was piped. Three years later the harbour trustees went out of business and the docks were taken over by a company set up by the Great Western Railway. Alexandra had been looking for work at ports other than Liverpool, in order to ride out the shipping depression of the early 1920s. They do not seem to have been working at Cardiff any more, but in November 1921, Procter was negotiating with the dock company that ran the coal-exporting Barry Docks near Cardiff; and in 1923 feelers were put out to Watkins's old London clients, the New Zealand Shipping Company. These approaches came to nothing; but in March 1924 the foundations of a new outport were laid. The *Canada* was stationed at Swansea, another tug followed in May, and the *Albert* and *Herculaneum* set off after them in December. A permanent presence was needed to develop local contacts and business, and by January 1925 an office had been opened with a manager, T.H. Richards, in charge. In a small, steady way it became established as

'Huskisson' with an unusual number of visitors on board.

In dry dock — showing the size and shape of the hull. 1000 h.p., 201 gross tons, powered by triple expansion machinery.

a source of revenue, yielding between £150 and £300 a week by late 1925. In the winter, Richards sometimes did better than Stark at Southampton.

Rivalry between Swansea tug owners was as fierce as ever, exacerbated now by the control over the docks exercised by the Great Western Railway,

'Nelson' built at J. Crichton & Co. Ltd, Saltney. 1,000 h.p. 201 gross tons - completed in January, 1935, was followed by a similar tug 'Salthouse' in March that year.

and the presence of two small GWR tugs. Competing with them from 1924, Dewsbury added the *Benson* to their little fleet. This tug had started life as the first Alexandra tug to be christened *Egerton,* and had passed into Admiralty hands. There were no friendly arrangements between competitors, as at Liverpool and Southampton.

To improve the chances of its own tugs, which in the words of Alexandra's manager were so underpowered that they 'couldn't pull a herring off a gridiron', the GWR made life as difficult as possible for the port's other tug operators. In summer 1932 a confrontation began. The dock company demanded a locking fee of £5 for every other tug in the port, and the right to engage tugs on behalf of shipowners, for which service they required a 2½ per cent commission on the tug's towing fee. The tugowners decided that they had no choice in the matter, since business at Swansea was good, so they paid the commission under protest. Later, it was raised to 5 per cent. However, Alexandra held a strong hand, in that the Elders & Fyffes fruit boats that coaled at Swansea insisted on using their tugs. When Alexandra threatened to withdraw, as they were being forced to work uneconomically, Elders & Fyffes put pressure on the GWR to moderate their demands. As a further inducement, Alexandra offered the railway company the use of one of their Swansea tugs at nearby Port Talbot. The £5 locking fee was withdrawn. The *Canada*, with a Swansea crew, was stationed at Port Talbot from 1937, although this was an unprofitable operation for Alexandra.

The GWR tried to work their own tugs at Port Talbot, but these were badly run, and a failure.

To add to their troubles, Swansea tugowners endured a running battle with their crews in the mid-1930s. A strike of all tugmen in the port began in December 1934 and lasted until the following March. Forty out of about seventy men came out, affecting the work of the five Alexandra, four Britannia, and two Dewsbury tugs then in the port, though not the two GWR tugs. The strikers' objective was recognition by the owners of the Transport & General Workers' Union, which they would achieve indirectly by extracting an agreed uniform scale of wages and conditions from all operators. The tugmasters stayed at work, and outside crews were brought in. Alexandra's revenue stayed constant. Tugs were sabotaged; nuts were found in the steering gear; engine room telegraphs were damaged. Dewsbury settled, and eventually the other owners followed suit, on the men's terms.

But Swansea shared in the revival of shipping that began in 1937. In that year nearly 7.4 million tons of cargo passing through the port - a higher

'Eastern City' being looked after by 'Coburg' which was built in 1934. She and 'Huskisson' cost £26,800 the pair.

A famous name is 'Flying Kestrel' in the Alexandra fleet. This is number four of the 'Kestrels'.

figure than that of 1913. It was exporting oil products, steel, and chemicals as well as coal and tinplate; it imported oil, metal ores, pit timber, grain, and other foodstuffs.

Port Talbot, Swansea's satellite seven miles to the east, had had a dock since 1837; but the fact that it was closer to the South Wales coalfield than either of its big neighbours - indeed, was actually on a coalfield of its own - was unexploited until 1894. In that year a dock and railway company was founded to develop the coal and port. By 1899 a new, 23-acre dock and a railway network to serve it had been built. By 1905, the dock area had grown to 75 acres; and by 1937 to 90 acres handling nearly 2.1 million tons of goods. Port Talbot shared Swansea's coal and tinplate export trade, and imported pit timber and metal ores. Here the Swansea tugs worked as well, and later on, Cardiff tugs too. Alexandra had first cast an appraising eye on Port Talbot as early as 1908, when the dock company had expressed an interest in the services of the company's tugs: in time they were to have the field to themselves.

From a low point in 1926, Alexandra's Mersey business was beginning to look up in 1927 - though too late, alas, for Cran, who were wound up in March. By then the battle with Rea had reached a draw. In the previous November it had been agreed that Alexandra were to have half the Merseyside dock business, and Rea and Cock Tugs a quarter each. Each would keep the customers they already had, and would be free to charge them what rates they could get. There would be no poaching. Other operators, including the Mersey Towing Company, became parties to the arrangement. Naturally, rates began to climb again: the shipowners' reaction is not recorded, but may be imagined. In February 1927, when recovery had begun, it was as if the long, bitter struggle had never been. At times when Alexandra had more work than they could handle, they were hiring boats from both Cock Tugs and Rea.

Weekly revenue climbed to over £3000 again. The opening of the rest of the Gladstone Dock was an omen, it was hoped, of better times. Certainly Alexandra lost no time in renting an office and a telephone there. Towage rates went up by 12½ per cent for Cunard and White Star. The second half-year brought in a revenue of nearly £81,000 - compared with £69,000 four years before - and a net profit of almost £5900. Similar figures were registered during 1928 and 1929. The Southampton office made a notable contribution to the happy picture; by mid-summer 1929 income from that direction frequently rose above £1000 a week, accounting for a third of total revenue, and it persisted at a high level even when bad times returned, making a doubly valuable contribution. Half-yearly dividends, interim dividends, and bonuses for shareholders and directors became routine once more. All tugs were busy, and there was now room for more. The *Herculaneum*, which had been sold, was bought back and renamed *Hornby* in June 1927. A new tug was ordered. This was the second *Sloyne*, to the same specifications as the *Wellington*, ordered from J. Crichton of Saltney and launched in August 1928. The old *Romsey* was sold, and was replaced in Southampton by a new 'tender

'Crosby', another of the tugs which attended 'Thetis' in Liverpool Bay in 1939.

adapted for towing' of the same name, built by Ferguson Bros of the Clyde. This 509-ton, twin screw boat, costing £33,750, was launched in January, 1930.

Early in 1930 the shadows began to gather again; and this time they stayed dark in most areas for seven long years. Booth, Anchor, and Brocklebank asked for rate reductions. So did Cunard, Canadian Pacific, Royal Mail, Union Castle, and White Star, for not even the Southampton operation, heavily reliant on passenger business from depression-wracked America, was quite exempt from bad times. Alexandra resisted; but early in 1933 their Liverpool rates came down by 5 per cent. White Star were in deep trouble. They were late paying their accounts in 1930, and again in 1931. They were taken over by Lord Kylsant's Royal Mail group of companies, which was over-extended and in 1933 failed in the conditions of depression. Finally, in 1934 they were merged with Cunard to form a single, more competitive business.

Liverpool's income was the main sufferer. Turnover that had stood at close to £2000 a week dropped below £1500. Worse was to come: there were weeks early in 1932, and again in 1933, when the company's total revenue from all sources was hardly more than that figure. This was the blackest hour of a black time. The company stayed in profit, but it was depleted. The first half of 1930 showed a figure half that of two years earlier. Revenue for the second half of 1932 was back to below the 1925 figure.

As business dried up, so tugs were laid up once more - seven in June 1931, five in November, including one each at Swansea and Southampton. There was no tug building this time. Additions to the fleet were made on the cheap, and were confined to Southampton. Alexandra worked the North German Lloyd tender *Grúśsgott* under an anglicized name, *Greetings*, towing the visiting ships of that company and of the Hamburg-Amerika line. In addition they were allowed to seek what other business they could, and keep a quarter of the revenue from that source. Since all operating expenses were paid by North German Lloyd, the arrangement was clear gain. And in January 1933, Alexandra bought the antique White Star tender *Magnetic* for £3000, converted her to oil firing, renamed her *Ryde*, and set her to work for the Southampton office.

The makeup and character of the board changed completely between 1930 and 1935. Barry Bicket died in December 1930, at only thirty-eight. His brother Alexander Bicket Jr (known as Alex), a marine insurance broker who handled the Alexandra tug business, filled the vacancy two months later; but their father Sir Alexander also died in March 1931, aged seventy-eight. In July 1933 J. J. Mack died at the age of seventy-seven, and Alex Bicket was appointed Chairman in his place in August when Silvester Procter, Charles Garrett Mack, a son of Isaac, and Harold Brooke Mack, a son of J.J., joined the board.

Clear view of the Voith Schneider propellers and their protecting skegs.

Alexandra tug attending a submarine in the Thames. A Canadian flag just visible at her stern. (Handford Photography, Croydon).

'Canada', a Voith Schneider tractor tug, entering Royal Seaforth Dock, was delivered in 1980 by McTay Marine, the first vessel built for the Alexandra fleet in that yard. (P.B. Parker, Formby).

'Collingwood', 1981, with one member of the crew about to get very wet!

'Avenger' with a long line out to her tow. (Fotoflite, Ashford).

'Deben' off Europa Point, Gibraltar.

Built in 1987 at McTay Marine Ltd, Bromborough, 'Waterloo' is seen here on acceptance trials passing the famous river front of Liverpool.

The 495-ton 'Invincible', built in Norway, joined the fleet in 1983. She was the most powerful of the vessels at that time with a bollard pull of 100 tons. She is seen here with a diversified deck cargo.

'Wellington' towing in Liverpool Bay. Photograph taken from a Britten, Norman Islander owned by Alexandra Aviation.

'Sun Thames' built by McTay Marine Ltd in 1982. Voith Schneider propulsion with a bollard pull of 30 tonnes and full fire fighting capability on board. (Hall Mullin, Bromborough).

'Irwell Trader' one of the vessels owned by Bulk Cargo Handling Services alongside the grain loading facility at Royal Seaforth. (P.B. Parker, Formby).

'Traveller' transporting Conoco's three wellhead decks destined for the South Valiant, Vulcan and Vanguard field.

AMT 'Traveller' with load out of the Statoil module No. 24M at Egersund for the Gullfaks C Project, Norway, February 1988.

Euroair show their colours in three different types of aircraft.

'Canberra' returning from the Falkland Islands to Southampton on the 11th July, 1982, with 'Brockenhurst', one of the tugs in attendance.

Unusual job for 'Indomitable' which went 600 miles out into the North Atlantic in a full gale to refuel Richard Branson's 'Spirit of Britain' the 'Virgin Trans-Atlantic Challenger' 1985, which sadly broke up and sank within the grasp of the record.

Fine view of 'Ark Royal' during her visit to the Thames in 1987 with the Royal Naval College, Greenwich, in the background. (Chorley & Handford).

'Flying' sisters — the 'Kestrel' and 'Osprey', two of the present fleet of Alexandra tugs at Southampton.

'Invincible' acting as support vessel for the Admiralty Research Establishment (ARE) trials in Norway in 1987, with a high speed buoyancy propelled vehicle known as Moby-D.

Alexandra tugs and barges at Coryton on the Thames working for M.W. Kellogg at Mobil Oil installation.

In 1989 'Invincible' embarked on the longest tow in the history of the Company. Seen here off Capetown she towed the submersible barge 'Goliath Atlantic' from the Gulf of Mexico to Singapore, then to Yokohama and on to California, a total of 21,000 miles.

ITM 'Mariner' with 'Redoubtable' and 'Waterloo'. The crane is capable of lifting 200 tons and there are workshops, hospital and accommodation for 90 men.

Alexandra tugs attending the Queen Elizabeth 2 on her first visit to Liverpool in celebration of Cunard's 150th Anniversary, 24 July 1990 (B. S. Clarke).

The liner R.M.S. 'Queen Elizabeth' about to depart from Southampton on her last westward crossing on 23rd October 1968. 'Brockenhurst' in attendance (R. Bruce Grice, Fordingbridge).

R.M.S. 'Queen Elizabeth' being berthed stern first in October 1954 alongside the Ocean Terminal, Southampton. This was the first time a large Atlantic liner had been berthed in this way since 1922.

During 1934, Frederick Poulsom died and Charles Kerr joined the board; although Kerr, too, was dead by July of the following year. This was the same Charles Kerr who had been abruptly unseated from the board forty years before: he had remained a substantial shareholder. Ill-health forced Procter to retire as company secretary in September 1935, and he did not live long enough to enjoy the generous pension bestowed on him, dying just two months later.

As in the 1890s, we see a deliberately risk-taking policy introduced, of building up at great expense (in spite of poor revenue) a fleet of new tugs to take advantage of better times when they came - for otherwise, the business would go to operators who could provide modern boats. For this to be possible, costs had to be cut - the second leg on which the new programme stood. In 1934 the average age of Alexandra's tugs was thirty years. The best and easiest way to reduce costs, given the growing burden of repair bills and insurance premiums, was to sell the older boats. Such a policy went hand in hand with building up a new fleet.

No fewer than ten boats were sold in four years from mid-1933: the *Hodgson, Coburg, Alfred, Huskisson, Wapping, Nelson, Hornby, Ryde, Albert* and *Toxteth*. The last two were tainted with tragedy. In January 1933 a sunken coal elevator in Canada Dock had torn the bottom out of the *Albert*, which went down in seconds, taking three men with her. The *Toxteth* capsized in the Brunswick Dock in February 1937, drowning four of her crew. She was raised, but was immediately sold.

Eight tugs were built to replace those sold. A new *Coburg* and *Huskisson* came from Crichton early in 1934, built cheaply - £26,800 the pair - while prices were low from lack of demand. With their triple expansion, 1000 hp engines, they were the best of their kind. A new *Nelson* on the same lines was ordered later in the year, and a new *Salthouse* too; both from Crichton. Cochrane & Sons of Selby in Yorkshire got the contracts for the next four tugs; the new *Hornby* and *Wapping*, delivered in September 1936, and the *Alfred* and *Crosby*, the first of her name, in service in summer 1937. The last two boats cost £15,520 each.

For these four years, it was touch and go, for in spite of economies, costs trended upwards. Apart from the building programme, there was the generally rising price of coal. It was 16s a ton in 1934, 18s in 1936, 20s in 1937. Until business improved, towage rates could not be increased to compensate. In Liverpool, Alexandra's situation was not helped when the Mersey Towing Company pulled out of the trade-sharing scheme and started to compete. Overdrafts were often needed, and net profits sank even further. A dividend of 5s a share was usually paid, which by the beginning of 1935 was causing unrest among share-holders.

Then, in the second six months of 1935, the first fruits of good housekeeping appeared - the net profit figure was almost £4000. A bonus of 1s 3d a share was paid for the first time in years, silencing the murmurs. The revival of the shipping business in 1937 brought the anticipated boost to Alexandra's revenue, setting the seal on their recovery. It gave Alex Bicket great satisfaction to report, at the fiftieth annual general meeting of the company, that the revenue figure for the past half-year was the best for nearly a decade. Liverpool dock and river income for the past three months was up 17 per cent on the equivalent period for 1936.

In a smaller way Cock Tugs, too, had taken the calculated risk, building in bad times as an investment in better times to follow. They built three new, powerful, general-purpose tugs, all 95-footers from Cammell Laird, with triple expansion engines developing 1000 hp - the *Grebe Cock* of 1935, the *Storm Cock* of 1936, and the *Black Cock* of 1939.

Alexandra's achievements were the greater, in view of the inexorably escalating costs. The rise in revenue in 1937 over 1936 was more than swallowed up by an increase in expenditure on coal and wages. The increase in the wage bill arose from recognition of the Transport & General Workers' Union, after nine years of suspended relations, which meant paying union minimum rates. These extra costs caused Alexandra to raise its towage rates by 10 per cent. As usual, some clients refused to pay, while others, including Cunard-White Star, acquiesced. Costs rose again in 1938, but it was still a 'splendid' second half-year for Alexandra. Both revenue and the company's image were boosted by such work as the launch of the Cunarders *Queen Mary* in 1934 and *Queen Elizabeth* in 1938, from John Brown's yard at Clydebank. On each occasion Alexandra tugs from Southampton assisted, and two Alexandra tugs attended the *Queen Mary* on her trials in 1936.

In the first week of June 1939, Alexandra got one job that they would rather have done without. The submarine *Thetis,* on her trials in Liverpool Bay, flooded during a test dive with fifty-three crew and fifty passengers on board. She ended up nose-first on the seabed 160 ft down, with her stern protruding 18 ft above the surface. No one knew where she was, for her wireless was put out of action, and the *Grebe Cock,* which had kept her company, had left her before she dived to make for her expected surfacing point. Rescuers did not find her for eighteen hours, by which time the air on board was just about exhausted. Only four men got out, through the escape hatches.

During the second week of June rescue craft including the *Crosby* and *Grebe Cock* arrived with divers and camels - lifting vessels. By this time no one on board was left alive, but the rescuers could not be sure of this. Hampered by strong tides, desperate efforts were made to raise the submarine. Divers put a wire around her, and this was attached to a camel. The *Thetis* was rising at a steep angle when the wind blew up, putting extra strain on the wire. The surviving crewmen of the *Crosby* would never forget the shower of sparks as the wire parted, sending the submarine back to the bottom. Again a wire was put around her; this time it slipped off as the *Thetis* slewed.

The camel tipped up, and for the third and last time the *Thetis* plunged. In all, the *Crosby* stayed on station for five days, while excursion boats full of sensation-seekers hovered. But it was useless. Nothing now remained but the melancholy business of salvage, five months later.

When Alex Bicket presented his first half-year's report for 1939, the prospect looked set fair; increased revenue and, for a change, reduced costs. But in the same month, Alexandra tugs set sail once more for the naval bases at Milford Haven and Devonport, on Admiralty requisition.

10

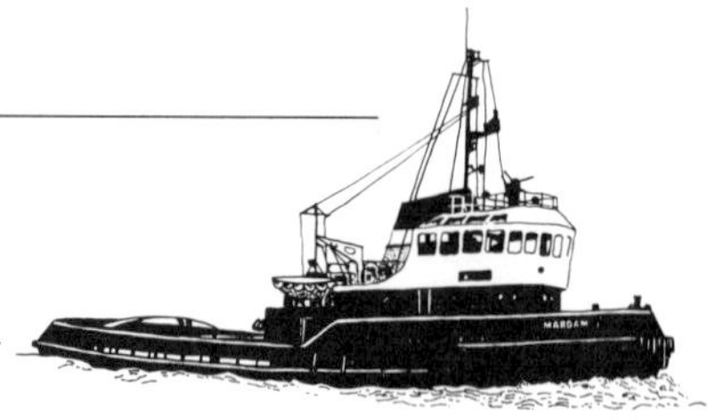

Margam

War Service: II

For the tugs, the Second World War was very different from the First. By June 1940, after only eight months of hostilities, German forces were occupying all the continental ports facing Britain. There was none of the orderly to and fro of convoys across the Channel, with men and supplies for a war front still on French soil: these narrow seas were no longer a corridor but a battleground, dominated for long periods by the Luftwaffe. The assembly and reception of convoys in Britain's ports and rivers were the over-riding priority, and with the development of long-range bomber aircraft, these ports were no longer havens of safety, but beleaguered front-line targets.

The number of civilian tugs requisitioned by the Admiralty as examining vessels or for towing casualties was only a quarter as great as in 1914-18, for after the fall of France and Belgium there was greater need for them in convoy work. Only seven Alexandra tugs are recorded as commandeered, and at the war's end the number still away on naval service was five. Nor did requisitioned tugs sail the world, as their predecessors had done. Alexandra's *Flying Kestrel* served at Suez between 1942 and 1946, but she was an exception.

The same could not be said of the workforce, many of whom joined the Royal Navy. One was on board the destroyer *Hardy*, on the way back from escorting a convoy to Russia when she was torpedoed. He managed to jump aboard another ship as his went down, but forty other men were not so lucky. In 1944 he was on a landing ship coming out of Patras harbour in Greece when she was mined. On her way to Alexandria for repairs, the ship lost her escort in a gale, then ran ashore on a sandbank on the North African coast. After several days she was pulled off - by a tug.

Most tugmen fought their war closer to home, in armed trawlers fitted out for escort work and minesweeping. Such little craft as these bore the brunt of the war in the narrow seas. Escorting colliers and coasters, a trawler

'Sun' tugs in the sunset at Woolwich, 1937 (R.E. Batten, Manor Park).

out of Lowestoft was bombed in high-level attacks and machine-gunned in low-level ones. The wooden lifeboat took one bomb through its bottom; another bomb lifted the trawler bodily out of the water. Much of the work consisted of picking up casualties, choking on oil and scarred by exposure; among them children, schoolteachers, and nuns from the *Simon Bolivar,* sunk off the North Foreland in 1940. The minesweeping winch on board the trawler was useful for towing ship casualties, with an axe handy to cut the rope if a raider appeared. Later, the boat was attacked in the Irish Sea by a submarine, all of whose torpedoes missed.

The crews of the tugs called upon to work outside their home ports had an equally hair-raising time. One recalls sweeping bomb fragments off the deck of the *Gladstone* while she was being dive-bombed at Devonport. Harold Russell was master of the *Kenia,* a requisitioned Watkins tug that had been armed with machine-guns and furnished with Navy gunners and signallers to act as a rescue tug. She went out to five ships ashore near the Sunk light, and had pulled off two when a German bomber pounced, hitting both tows with bombs and machine-gunning the *Kenia*. Bullets riddled the wheelhouse, holing the compass and speaking tube. One passed through Russell's cap band, scoring his face so that it swelled up like an egg. On another occasion the *Kenia* was trying to salvage an Italian ship on the Goodwin Sands. She hit a mine and blew up just as the *Kenia* was getting a rope on board. On the tug, hull plates parted, bunks and galley shelves collapsed, and debris showered everywhere. In 1939 the United Towing Company of Hull owned three of Britain's five ocean-going tugs, and two of these were sunk. It must have afforded some satisfaction when their *Seaman* shot down a German bomber.

Tugs really came into their own in the evacuation of the Allied armies from Dunkirk in 1940, for they were ideally suited to towing troop-carrying barges and disabled vessels, of which there were hundreds in the armada of large and small craft that took part; and with their manoeuvrability, they were adept at dodging bombs and low-level machine-gun attacks in confined spaces. Every Thames tugowner sent his boats. Six of Alexander's 'Sun' tugs were among them; in all, these six brought home more than 1400 men.

Few were as busy as Sun *XV*. On 30 May she towed six lifeboats to the beaches at La Panne, east of Dunkirk, under constant harassment from hostile aircraft. When fifty yards offshore she sent her boats in, now under intense artillery fire as well. They delivered seventy or eighty soldiers on board before all were sunk. Still under intermittent machine-gun attack, the tug carried her human cargo safely to Ramsgate; then went out again on 1 June to rescue two disabled and helpless small craft full of troops in the Channel. She got a rope aboard a bombed and empty hospital ship, but was ordered to abandon her when machine-gunned by a German aircraft. Instead, the *Sun XV* found a broken-down drifter laden with soldiers and towed her home. She returned to Dunkirk on 3 June with four boats in tow, but found no one left to take off. Homeward bound under repeated attack, she picked up a broken-down drifter.

Gaselee lost the *Fossa*, beached at Dunkirk; but firms with many more tugs committed were astonishingly lucky. Alexander was one; Watkins another. With at least seven boats at the besieged beaches and piers of Dunkirk, and two or more on towing duty at Dover, helping ships in and out of port, Watkins lost none; although all round them ships burned, blew up, and foundered. William Simmons's tug *Cervia* towed barges across, taking on some soldiers from the beaches and rescuing others from machine-gunning in the water, in all bringing 270 men back to Ramsgate. Once, a stick of bombs fell so close that the *Cervia*, like the Lowestoft trawler, was blown into the air. The *Java* worked at Dunkirk for the full three days of the evacuation, picking up soldiers and downed aircrews, as did the *Tanga*. Russell's *Kenia* in her rescue role searched the Channel for broken-down or damaged ships. The *Persia* towed home the crippled destroyer *Ivanhoe*, while the *Vincia* rescued survivors from the destroyer *Keith*, which was wrecked by bombs.

The luck of the Thames tugs was bound to run out, for even back in their home river, they were in the front line of battle, and were committed to five years of war. Considering what they endured, it is surprising that so many survived. Between 1941 and 1944, the Port of London Authority formed the London tugs into a standing fire patrol of seven or eight boats from different owners. In the docks, their crews routinely boarded blazing

Alexandra tugs at war. 'Crosby' and 'Alfred' towing an anti-aircraft fort to the mouth of the River Mersey.

Scott & Sons, Bowling converted a British Petroleum fire tug and she was given a famous old name, 'Flying Breeze' (A.T. Kelly & Co. Ltd., Glasgow).

oil barges and ships to put out fires. and towed others to safety. Gaselee's *Musca* and *Vespa* became naval auxiliaries, working as armed minesweepers. One of their tugs was lost in the Swale. Alexander's *Naja* was destroyed by a V1 flying bomb in 1944, all six aboard being killed. Their *Gnat*, which, ironically enough, had already been run down and sunk by a German ship in peacetime, now took a direct hit from a bomb. Alexander's *Sun VIII* and Sun *IX* were both sunk by mines in the Thames estuary. The Watkins fleet was badly hit too. The *Napia* was blown up by a mine with all her crew; the *Muria* was lost with all hands off the North Foreland; the *Persia* was wiped out horrifically when a loaded tanker hit a mine at Thames Haven, swamping tug and jetty in burning oil. More than sixty died.

After the fall of France Britain's first priority became the preservation of the convoy system, plugged into the nation's life support system in the Mersey and the Clyde. In July and August 1940 the *Wellington, William Poulsom, Flying Breeze, Flying Kestrel* and *Sloyne* at Southampton, together with the requisitioned tugs at Devonport and Pembroke Dock, were pulled back to Liverpool to help handle conveys, while the *Romsey* replaced the boats at Pembroke Dock. For the *Romsey*, it was out of the frying pan into the fire. She was mined in January 1941 off Milford Haven, repaired, then in September 1942 run down and sunk in the Clyde with the loss of fourteen men. She was subsequently raised. The Red Funnel tugs took punishment too. Their *Sir Bevois* was bombed at Plymouth in March 1941. The *Canute* was sunk at her Southampton berth in December 1940, but salvaged. As for Swansea, the centre of the town was obliterated. The windows of Alexandra's offices were blown in, and the Britannia tug *Queenforth* was sunk at her moorings. After long hours of convoy work and no time off, the morale of the tugmen momentarily cracked. They walked off their boats, and had to be turned back by police.

But of all the ports that had escaped the immediate horrors of war in 1914-18, it was Liverpool that suffered the most now, while bearing the biggest responsibility for keeping Britain supplied. Between the end of July 1940 and January 1942, there were sixty-eight air raids. Ninety-one merchant ships were sunk or badly damaged in the docks; out of a total of 144 berths, a third were put out of action. Of the warehouses, nearly half were destroyed. During the worst of the blitz in May 1941, the Brocklebank steamer *Malakand*, with 1000 tons of bombs and shells on board, was set on fire and blew up. Sheds and quays simply disappeared. The ship's plates were found two miles away; her four-ton anchor was blown 100 yards. Amazingly, only four lives were lost. Yet in the six years of war, the port received 1300 convoys of up to sixty ships each, and handled 75 million tons of cargo. The tonnage of ships turned round during 1940 sank to 2.3 million tons, compared with 21.7 million tons in 1939; but it climbed to 6 million tons in 1942, in spite of the destruction, and the 1945 total was the same as for 1939.

Tugs, tugowners, and tugmen were naturally in the thick of it, while still having to work around the clock. Bombs and aerial mines wrecked homes, offices, and tugs. The offices of Cock Tugs were set on fire and gutted; those of Lamey and Alexandra were heavily damaged. Serious attacks began in December 1940; in March 1941 there was another major assault. The *Crosby* was hit by an incendiary bomb; the *Nelson* caught fire; the *Wellington* was badly damaged when a bombed warehouse wall collapsed on her, putting her into dry dock for six months. This was a hazard: the *Coburg* later suffered in the same way. Just before Christmas came another blitz, in which the *Alexandra* was set on fire. The *Collingwood* was damaged by a bomb in March.

In the first week of May 1941 the *Alexandra* and other tugs were truly tested. The *Hornby* was sunk, though afterwards raised; the *Alexandra, Canning*, and *Brunswick* were all damaged. It was a miracle that no tugmen were killed. One master remembers lying alongside a barge full of burning timber while an ammunition ship exploded astern of him, and shrapnel rattled down all around. Other tugmen were below decks in the *Alfred* in the Gladstone Dock, playing cards, when a bomb hit the dock. The blast blew the hatches off, leaving the card players sitting in the open air, and bodily shifted the wheelhouse of the nearby *Sloyne*. A bomb fell so close to the *Alfred* when she was in the river that she was swamped. Once when working in the Irish Sea, she was set upon by hostile aircraft but escaped damage. Every morning, before the day's docking and undocking began, the Mersey had to be swept clear of mines. Even so, Rea's tug *Poolgarth* hit a mine off the Canada Dock and was blown to pieces; no trace of her crew was ever found.

In June 1944, Britain's tugs took sweet revenge for the years of punishment and retreat. From ports all around the coast, 132 boats assembled on the Solent to help the Allied armies return across the Channel. Alexandra tugs had come back from Liverpool; Lamey tugs joined them; Crosthwaite tugs came from the Tees; Batey tugs from the Tyne; United tugs from the Humber; Gaselee tugs from the Thames. They had brought with them, all or part of the way, concrete caissons to make up the prefabricated Mulberry Harbours that would create instant ports on the hostile shore; and sections of PLUTO - pipeline under the ocean - that would supply the invasion armies with fuel.

Meanwhile, Alexandra's difficulties did not end with damaged tugs and wrecked offices. In this war there was an absolute ban on the building of new tugs, and the government strictly controlled the overhaul of existing boats. Costs soared, as in the First World War. Insurance premiums rocketed, to cover war risks. The price of coal, already 29s a ton in December 1940, rose to 40s and more in 1944; between 1939 and that year, expenditure on coal went up by nearly 120 per cent. Wage increases at all ports in 1940 were followed in 1941 by rises of between 15 and 25 per cent, and in 1943 and 1944 wages went up again. In addition, the men got overtime and an extra payment for war risks. As a result the wage bill rose by almost 94 per cent in the same five-year period.

Consequently, towage rates went up, once in 1939 and twice during 1940. Revenue at Southampton declined and finally vanished when Alexandra's tugs were withdrawn; but elsewhere it rose steadily. The increase was particularly noticeable in Liverpool, where - in contrast with the First World War - the volume of shipping using the port rose after the initial drop, instead of declining. The work Alexandra was doing doubled between the outbreak of war and mid-1941, in spite of the blitz. Total weekly revenue went up from around £4000 late in 1939 to more than twice that figure early in 1945.

These weekly revenue figures show that not even the most strenuous efforts of the Luftwaffe could stop movement in Liverpool for long. For a fortnight during the blitz at the end of August 1940, and again at the beginning of May 1941, Alexandra's dock and river revenue fell by 20 per cent; but the drops were so undramatic, and so typical of normal fluctuations,

'*Empress of Britain*' with '*Canning*' in charge (F. Leonard Jackson, Heswall).

that they cannot be positively attributed to the bombing. In Swansea the effect was clearer-cut. In the 1940 attack Alexandra's revenue fell by 40 per cent for two weeks, and in the corresponding period in 1941, none at all was recorded.

Thereafter, the company's income suffered no more major setbacks. Helped by the return to Southampton in May 1944, half-yearly payments to shareholders settled down at 6s 3d and a 5s bonus per share. Net profits per half-year averaged nearly £7500. At the end of 1943, the 32-tug fleet was valued at £564,000, in spite of the battering it had taken. Even the much-bruised *Romsey* was worth £36,000.

But as in the First World War, Alexandra and their directors saw only a small proportion of the increase in profits. Most of the transfers to reserves were earmarked for tax. An Excess Profits Tax was reinstituted, this time taking 100 per cent of the profits above a standard rate; and income tax rose to 8s 6d and then 10s in the pound. There was so little leeway that when, early in 1943, there came a demand from the Inland Revenue for £77,000 in income tax and Excess Profits Tax, Alex Bicket found £25,000 out of his own pocket and the bank provided a temporary overdraft. Furthermore, the company, motivated by patriotism, waived the interest payments due on their government stocks, and on one occasion paid some income tax and Excess Profits Tax in advance, foregoing the use of the money for that period. However, the description 'very gratifying', applied to one half-year's operations, could have been applied to most; and the company celebrated VJ Day in August 1945 with assets battered but in sound financial shape.

11

Redoubtable

A Changing World

The tugs that went to war in 1939 were not so very different from their predecessors of 1914 - indeed, they were often the same boats. Because its work did not radically change, the typical dock and river tug of the early years of the century was still at home in the same waters even later, into the 1950s. On Merseyside, for example, there was the Cock tug *North Cock*, built in 1903, a handy 96 ft for manoeuvrability, with a compound engine developing 700 hp, and capable of 12 knots running free. Such tugs might lack the power of more modern boats, with their 1000 hp triple expansion engines, but given more time to do the job, they could still happily handle the same tow. Britain's most advanced sea-going tug of 1914 was the 1800 hp *Hullman*, owned by the United Towing Company of Hull, as much at home in the open ocean of the 1950s as she had been in 1914.

The steam tug had reached its peak of development. Such changes as were made to it were of detail, improving its efficiency. Perhaps the most important was in the area of communications. Wireless telegraphy, using Morse code, established communication between a tug at sea and its home port or other ships for the first time, enabling orders, and changes of orders, to be sent without need for the boat to put into port to receive them. Tugs that had been fitted with wireless while on Admiralty service during the First World War sometimes retained it when they returned to civilian service - Alexandra's *Gladstone* was a case in point - but tugs with wireless remained in a minority. Morse communication was a leisurely and skilled business, ill-suited to the controlled confusion and instant decision-making of dock and river, where most tugs worked. The bigger and more powerful tugs, such as Watkins's *Kenia*, that were the most likely to be sent on the longer-drawn-out jobs such as salvage, were given wireless first. Dock and river tugs continued to communicate with their tows by the time-honoured method of whistle and siren codes: the ship sounded its siren, which the tug acknowledged with its whistle.

When the new Elliott tug 'Vanquisher' arrived on the Thames with diesel engines in 1955 she was something of a rarity (M.J. Gaston, Dunton Green).

The radio telephone - more flexible, quicker, surer, and easier to use than wireless telegraphy - had wider applications and spread faster. It was in use for ship-to-shore communication by the late 1920s; the Crosthwaite tug *Queen's Cross* had it in 1933, the Alexandra tug *Brockenhurst* six years later. The radio telephone works in the short wave band, at very high frequencies (VHF) making speech easy to beam accurately and to pick up, and the equipment compact. At the highest frequencies of the VHF range, the radio telephone is an ideal tool for short-range communication in dock and river between tug and shore and tug and tow. At the lower frequencies of VHF, speech can be received well out to sea. The VHF radio telephone enabled tugs to be deployed more economically than ever before: a message could place any tug, whatever its task, in the right place at the right time.

This unique photograph shows a Cock tug being girted. It was taken at the exact moment that the tow rope parted — which saved the tug from being dragged underwater. Members of the crew can be seen taking cover from such a situation.

RT was fitted to two Rea tugs and eight more Alexandra boats in 1947, by which time it was widespread. In February 1948 its value was illustrated dramatically when William Harvey Lamey, eldest son of the founder and current head of the firm, witnessed the sinking of the steamer *Baltic Queen* from the garden of his Merseyside home. His *J. H. Lamey* had had RT since 1946, so he was able to call her up at once, and send her to pick up survivors.

As far as the mechanics of the tug itself were concerned, the period was one of experiment rather than progress. Until the mid-1950s the diesel or motor tug made little headway outside the world of the river tug. Harland

& Wolff had diesel-powered launch tugs by 1920; five years later, United Towing of Hull put their *Motorman* into service. The six other new tugs in their fleet were steamers.

After studying diesel propulsion in America, where it was more common, William Crosthwaite built his diesel-electric tug *Acklam Cross* in 1933. Propulsion was by an electric motor, power for which was generated by a diesel engine. The electric motor, like the steam engine, could produce high power at low engine speeds, and could reverse the direction of its rotation quickly, making it more flexible in operation than the pure diesel of the day. The *Acklam Cross* had a battery-powered electric starter, which gave quick starting from cold, and her captain controlled his engines direct from his bridge, instead of through an engine-room telegraph. This reduced the number of crew needed below, and the whole complement numbered no more than four, making for economy of operation. But at £30,000 the *Acklam Cross* was twice the price of a conventional steam tug and could not pay, particularly in depression conditions. The same may have applied to the diesel-electric lighter tug *Lectro*, launched on the Thames by the Union Lighterage Company in the same year of 1933. None of these boats was more than an isolated experiment in the midst of a steam fleet. Gaselee with their *Agama* were towing ships on the Thames in 1938, but as late as 1955, when the new Elliott tug *Vanquisher* arrived on the river, diesel ship tugs were still a rarity there. On the Mersey, there was none; though since 1940, Lamey had had them on the Manchester Ship Canal, where towing was in a straight line. The day of the motor tug was yet to come.

From the 1920s there were also experiments with oil firing of steam tugs, instead of coal, which seemed to promise the best of both the diesel and steam worlds. Since oil produced more power for its weight than coal, gave more precise control over furnace heat, and (like the diesel) needed fewer crew below, it offered the diesel's economy of running coupled with the flexibility of response of the steam engine. Furthermore, because oil took up less space than coal for the same weight of fuel, it gave greater range. But oil firing failed to take hold until the 1950s, and by that time the diesel tug had overcome its initial disadvantages. The first British oil-fired tugs were Alexandra' converted *Romsey* of 1922, and some lighter tugs introduced on the Thames a few years later, one by the Union Lighterage Company.

Another experiment, of more significance in the long run, was tried in 1934 when the Falmouth tug *John Hamilton* acquired a Kort nozzle. In 1936 the Cock tug *May Cock* was similarly equipped, as was the J. P. Knight boat *Knighton* a year later. The first British tug to be given a Kort nozzle when built was launched on the Forth in 1937. Invented by the German Ludwig Kort to reduce wash in canal boats, the nozzle was an open-ended collar or tube around the propeller that concentrated the water stream produced. The effect of this was to conserve otherwise dissipated energy, giving greater

To avoid the danger of 'girting' a gog rope was (and is) used to bring the pulling point further aft (Stewart Bale Ltd., Liverpool).

In 1968 'British Admiral' of 106,278 D.W.T. was the largest tanker to sail up the Mersey. She required six tugs for undocking. (Elsams Ltd., Liverpool).

thrust at low tug speeds and increasing pulling power by 37 per cent or more. Alternatively, the same pull could be obtained with less power, making for economy. The disadvantages of the Kort nozzle were its high cost, and the fact that it tended to keep the tug in a straight line, countering the effect of the rudder. It was not, therefore, widely adopted in its original form.

In most ways, the mechanics of towing changed little. Some time after 1900 the gog rope put in an appearance, making life safer for tugmen. A tug pulling a ship's stern around was in danger of being girted, when the towrope, attached to a towing hook nearly amidships, assumed a right-angle to the tow. The gog rope was a bridle that anchored the towrope at the tug's stern. If the tow then 'took charge', it should simply pull the tug along stern first, not over on to its side. In practice, however, the gog rope was not infallible: the Alexandra tug *Toxteth* was fatally girted in 1937 when her gog rope broke.

By the mid-1950s the manila-and-wire towrope - 12-inch manila spliced to 3½-inch wire - was giving way to the wire rope for sea towing, though not for ship towing in harbour, or for barge work. One reason, in sea-going tugs, was the spread of the towing winch, which used wire rope because manila was too bulky to be wound on a drum. The powered towing winch automatically pays out rope when a predetermined strain is put on it, and takes it in when the strain is relaxed, so maintaining an even pull and avoiding sudden jerks. This is of great value in ocean towing. When a fixed towing hook was used, the rope had to be long and slack to counteract pitching. It was vital to maintain the correct tension, so a permanent, exhausting watch had to be kept, and constant changes of speed made. A moment's inattention or a slow response might easily result in a parted rope and a crewman dead or maimed. A fixed towing hook could be sprung to allow it some flexibility, or tripped by a quick-release mechanism on the bridge; but the towing winch was the most complete answer.

The winch, first adopted in America, found its way into British ports in the First World War in Admiralty tugs which, specializing in the rescue of damaged ships at sea, had a particular need for the device. When these tugs passed into civilian service after the war, their towing winches went with them, as on Watkins's *Rumania*. The winch was never found at this time on dock and river tugs, where a fixed hook provided the sensitivity of touch needed when twitching a ship delicately to and fro on a short rope.

Besides this modern, or reasonably modern, technology we find J. H. Lamey supporting century-old technology, when he bought the paddle tug *Troon* in the same year as the *John Hamilton* set out to work. A year later, Crosthwaite's fleet of eight tugs still included two paddlers for dock work. Nothing illustrates better the fact that this was, in most respects, a period of technical stagnation.

The work that tugs did in peacetime changed, in most ways, as little as the tugs and their tows. As usual, sea towing and salvage provided most of the drama, and all of the big disappointments. In 1923 the men of the Crosthwaite tugs *Athlete, Liberia,* and *Dora Duncan* suffered, in the words of a newspaper report, forty-eight hours of 'sleepless peril' searching through a gale for a broken-down ship that had drifted from her last known position. The seas were so mountainous that the tugs could not steer head-on into them for fear of swamping; yet even so, they took on board so much green water that they had to put into Bridlington to be pumped out before resuming their hunt. It was fruitless in the end, for other tugs got the prize.

This was a story that could have been told at any time since the mid-nineteenth century. Not so the lethal adventure of Alexander's *Sun IV* and *Sun XII* when the tanker *Elkhound,* laden with petrol, caught fire at an oil depot full of combustibles on the Thames in 1930. In a situation of hair-raising peril, in danger of being blown to bits at any moment, the tug crews grappled the blazing tanker and towed it away from the depot, while keeping the fire on board under control. By this time most Thames tugs had been equipped with pumps and monitors - water cannon - for fighting tanker and oil installation fires. The tugmen's heroism on this occasion probably saved the river from its greatest disaster since the Great Fire of London. The age of oil offered them an entirely new and horrific way of dying, with which they were to become all too familiar.

When peace came and international trade revived, shipowners found themselves in a sellers' market, as they has in 1919, and for much the same reasons. More than half of Britain's prewar merchant tonnage was at the bottom of the sea, and although it had been more than replaced during the war by American-built ships, these were standardized, basic cargo-carriers - mass-produced floating boxes. Half of Britain's most valuable tonnage - the modern, specialized, efficient cargo and passenger liners, and the tankers - had been lost and not replaced. The American 'Liberty ships' that abounded were suited only to tramping, not to the more lucrative trades. Furthermore they were built big - up to 10,000 tons - for maximum capacity, making them unsuitable for any but developed, roomy, deep-water ports. More than half of Britain's postwar fleet consisted of these vessels. Worldwide, with the virtual destruction of the German, Japanese, Italian, and French merchant marines, the total tonnage of merchant ships had fallen from 60 million to 35 million. War damage to shipyards slowed new construction: not until 1949 was as much new tonnage built as in 1938. Freight rates went up as a result, and shipowners did well. They built new as soon as they could, in spite of rising costs of construction, fuel, and labour. And, as ever, the tugowners rode on the back of the boom.

The difference was that, thanks to American economic support for the

The Cunard liner 'Queen Mary' arriving at Southampton from New York with Alexandra tugs in charge. (R. Bruce Grice, Fordingbridge).

The 'Queen Mary' leaving Western Docks, Southampton never to return, on October 1987. 'North Loch' takes the strain. (R. Bruce Grice, Fordingbridge).

war-ruined countries of the world, there was no swift collapse in the world economy as there had been in 1920-1; no catastrophic fall in freight rates because too many ships were chasing too little cargo. Because trade went on expanding, and war losses were still being replaced, freight rates continued to rise into the 1950s. Large-scale conflicts depending on sea-borne supplies helped - the Korean War of 1950-3, even the brief Suez War of 1956.

In Britain, the shipping boom showed signs of slackening from the mid-1950s. By 1955, Japan was building more ships than anyone else. The United States began to re-create her subsidized merchant marine of prewar days, concentrating on fast, specialized liners that creamed off the most valuable trade. American and Greek shipowners, attracted by the absence of taxes in Liberia and Panama, free to buy ships as cheaply as they could, and employing the cheapest of labour, built up fleets that enjoyed exemption from the usual commercial imperatives. By 1957, Britain's merchant fleet was third in size behind those of the United States and the 'flag of convenience' operators.

Growing world trade - expanding at 5 per cent per annum in the early 1960s - demanded more and more specialization in ship design. Its aim was fast turn-round, and economies in handling freight. The familiar shapes of the modern port began to appear, and the type of ship that tugs were called on to handle started to change. A new generation of bulk carriers was designed for the rapid loading and discharge of a variety of specialized cargoes, such as sugar, cement, metal ores, and methane gas. Loaded and unloaded by pump or conveyor, they were fast banishing the tramps, which lacked the means of taking on and discharging bulk cargoes quickly. A few tramps hung on, carrying grain and wool; but between 1967 and 1970, nearly nine out of ten new ships built were bulk carriers.

The age of cheap oil was the age of the tanker, the commonest of the bulk carriers, now accounting for nearly half of new world tonnage built each year. To take advantage of economies of scale, the bulk carrier was growing in size as well as numbers. In 1962, the 65,000-ton tanker *Olympic Challenger* was berthed at Thames Haven by three 'Sun' tugs; when in 1966 the 106,278-ton tanker *British Admiral* dry-docked in the Mersey, five Alexandra tugs were needed; and in 1981, the 339,000-ton ULCC (Ultra Large Crude Carrier) *Kirsten Maersk* dropped anchor at Coryton.

In the early 1960s cargo liners were still the commonest carriers of mixed specialized freight; but to allow them into smaller ports they were actually getting smaller, which was against the trend. On the American seaboard the container ship, expressly designed for the fast and cheap handling of such cargoes, had put in an appearance. Freight of various kinds, shapes, and sizes was packed into large boxes of uniform dimensions, stacked, if necessary, to well above the level of the rails. With specially designed cranes they could be quickly loaded and unloaded; and at their destinations the cargoes were not broken up but transferred straight to road or rail trucks for onward distribution. The cargo liner could not compete. By 1985, nearly 90 per cent of the freight that could be containerized, and that passed between ports equipped to handle the new boxes, was carried in this way.

The container ship had another characteristic, highly significant for ports and their tugs. Because it discharged on to trucks, it could unload at any port that had good road links, and where a container terminal was built for it. It was not obliged to use ports already at the hub of a rail network - that is, big, established ports such as Liverpool and London. There was positive encouragement for hitherto minor ports to develop by encouraging container traffic. In 1966 the American Sea-Land Company opened a container route to continental destinations, and also to Grangemouth on the Firth of Forth, a recently modernized port.

Britain's shipyards, the victims of foreign competition, labour unrest, and inordinately high costs, could not improve their increasingly outdated, expensive building techniques to meet the need for new ships. For the British shipping industry, a long slide began. From the early 1960s it was clear that it had too many ships of types that were less and less in demand. By 1962 freight rates were down to their 1948 level, while operating costs had more than doubled. Ships were laid up; and the full employment of the 1950s was over. The numbers of men in the industry began to fall, as did the number of ships for tugs to handle. From the early 1970s soaring oil prices brought economic crisis to the developed Western nations, reducing all kinds of economic activity, including trade. This again forced down freight rates, and demand for ships. The less efficient, less cost-effective fleets suffered worst, but ships of all sorts were laid up. Expensive oil caused tankers, in particular, to be broken up or 'mothballed'. The importance of this traffic for British ports may be judged from the fact that, even in 1982, at a time of recession, 65 per cent of shipping passage in and out was carrying oil or oil products. Container ships suffered least. In the lean years 1978-83, container traffic grew nearly six times faster than world trade as a whole.

Britain's merchant navy, ill-equipped to cope with hard times, fared correspondingly poorly. At its postwar peak in 1955, ships over 500 gross tons - the ocean-going fleet - had totalled 3575 vessels and 18.9 million tons. A long, fast slide in ship numbers then began - 2401 in 1965, 1682 in 1975, a mere 693 in 1985. Tonnage climbed at first as ships got bigger, peaking at 31½ million tons in 1975, but ten years later that figure, too, had collapsed, to 12.2 million tons.

When one particular type of ship went into decline, the Panamanian-registered bulk carrier could not, for once, be blamed. In the mid-1950s Britain found herself in possession of forty-six passenger liners. Many of these products of the early postwar boom were about to be turned into dinosaurs by the jet airliner, which made global travel quick and cheap for

The two 'Queens', 'Elizabeth' back in her traditional colours and 'Mary' in her wartime grey, Southampton 1946. (Southern Newspapers Plc., Southampton).

the first time. The 'floating hotel' concept could only be economic if fares were high. With escalating running costs, they rose steadily. In 1938, a one-way first-class passage across the Atlantic in the *Queen Mary* cost £57 10s. In 1946 the price was £90, which was already more than the air fare for a journey that took five days against fourteen hours. Then, in October 1958, the first transatlantic jet airliner services were inaugurated by BOAC and Pan American Airways. The crossing could now be made in a little over 8½ hours.

By the early 1960s more people were crossing the Atlantic by air than by ship, and both the *Queen Mary* and the *Queen Elizabeth* were running half-empty for more than half the year. A new sort of passenger liner emerged, less romantic but more economical to run. Designed not for the declining transatlantic trade but for cruise passengers and emigrants, it was smaller and simpler, and also more versatile in that it had more space for profitable cargoes such as perishable foodstuffs.

Other new ship types carried ambiguous messages for the towing industry. One was the semi-submersible heavy lift vessel, a self-propelled, sea-going pontoon that submerged its deck, on to which a ship or other floating structure could be manoeuvred. The water was then pumped out, raising the pontoon and leaving the load high and dry on board. It was less vulnerable to weather than a tug with its tow, and quicker. It did coastal tugs out of some long tows; but harbour tugs were still needed to manouevre loads on to it, as happened in 1986 when the polar exploration ship *Discovery* was transported from the Thames to her native Dundee.

The LASH (Lighter Aboard Ship) vessel was another mixed blessing. It did not need a port. It carried loaded lighters on board, discharging them at sea in the approaches to a port, or at the mouth of a river or canal too small for anything bigger than a lighter to negotiate. The lighters, towed by tugs, then delivered direct to the customer's waterside warehouse. The LASH ship was bad news for ports, whose docking charges, port dues, and delays it by-passed; and it was good news only for lighter tugs.

The foreign-built bulk carriers and container ships that called were big, and getting bigger. Container ships, in particular, were awkward to manoeuvre because of their tall, boxy superstructures and stacked deck cargo, which caught every breath of wind like a sail. Even the smaller ships might need a stern tug as well as one at the bows. Fast turn-round might mean more movements, and large ships had always needed more tugs; but speed and increasing size also signified that fewer ships were needed to carry the same amount of goods.

More threatening still was the increasing ability of ships both large and small to manoeuvre themselves in port at low speeds that normally rendered the rudder useless. This they accomplished with the bow and stern thruster, known since early in the century but only widespread from the 1960s, when

These views of 'Gower' show very clearly the two hooks and the quick release mechanism. This set-up was designed to prevent 'girting'. The hook swivelled round to port or starboard and the towing effort was not applied at the centreline of the tug but off to one side (Donald Cow, Swansea).

the growing size of ships made their advantages more apparent, and running costs - including towage - were subject to keener scrutiny. The commonest thruster unit consisted of a transverse pipe through the ship, at right angles to the hull, with a propeller in the middle allowing propulsion in either direction. Other types used jets, and some were rotatable through 360 degrees. All enabled a ship without steerage way to steer or move slowly astern without assistance.

If the new ships were to be attracted to a port - and so make work for tugs - port accommodation had to improve. Deep-water berths and modern, fast cargo handling were essential. Britain's established ports were slow, or ill-fitted, to react. One of the saddest cases was the Clyde. By 1960 the bigger bulk carriers, which had a draught of more than 40 ft, could no longer use the river. The largest able to do so were 28,000-tonners, which were by now in the middleweight league. Consequently new deep-water ports had to be built on the Firth of Clyde, at Hunterston for ore and coal carriers and at Wemyss Bay for tankers. The old Clyde withered.

But for the most part, the troubles of Britain's ports arose from unhappy labour relations. Where there had been modernization of cargo handling, the dockers were reluctant to accept the new systems, for fear of losing their jobs. Ships were delayed, losing money every day they were detained, so next trip they went elsewhere, increasingly to the great continental distribution centres of Rotterdam, Antwerp, and Hamburg. Destroyed in the war, they had been rebuilt from the ground up to give shippers the service they needed; they would work twenty-four hours a day, if need be, to fulfil a contract; and they were cheaper. As early as the mid-1950s continental ports were taking over the re-export trade of London and Liverpool. Even cargoes destined for Britain, such as American coal, were going first to the Continent, where they were broken up and shipped across the water by smaller 'feeder' ships.

The towing industry was, as usual, one of the first to feel the effects of such developments. It was faced by a problem recurrent since the shipping depression of the 1920s - too many tugs chasing too little business - and its reaction in the 1960s and 1970s was an equally familiar one: rationalization. If a port had work for ten tugs, it made better economic sense for them to join forces in some way than for, say, two operators with eight tugs each to compete. The result was a rash of new co-operation agreements and takeovers, but as times became harder still, from the mid-1970s, these tried and tested remedies became insufficient on their own, and tugowners were forced to seek new ones.

Helped along by economic and social pressures, the new demands on tugs forced to fruition, in the 1950s and 1960s, technical developments originating in the 1930s or earlier. The size and intractability of the latest generation of tows required a combination of unprecedented power, response, and

An improvement on the hooks on 'Gower'. 'Brockenhurst's' hooks took the strain at the casing side making sure the tug was pulled sideways instead of girting. (Walter Fussey & Son, Hessle).

S.S. 'United States' one of the many visitors to the new Ocean Terminal opened in Southampton in 1950.

manoeuvrability. The diesel engine in tugs had always possessed the potential advantage over steam that it was capable of much greater development: more power could be packed into a small space. By the 1930s the diesel was fulfilling its promise; powerful motor tugs were built for ocean towing, but only by the Dutch. There was no incentive in Britain to build very much more powerful harbour tugs than steam allowed until the 1950s, when the size of the tows compelled owners to make the change. The expense of building a completely new fleet was put off as long as possible, while existing steam tugs still had life left in them.

Increasingly, power was expressed not in terms of horsepower alone, but in bollard pull or static pull. This was the force, measured on a spring balance and expressed in tons, exerted by a tug pulling on a quayside mooring bollard. A much more meaningful measurement than horsepower, which could be interpreted in more than one way, bollard pull had been used as a standard, in addition to horsepower, since at least the 1920s. Very roughly, it could be said that 150 indicated horsepower, or 100 ihp with a Kort nozzle, was the equivalent of one ton of bollard pull.

At least as important for the tug as gaining power, it had lost the inflexibility that had kept the first diesel boats towing in a straight line. Gearing was one way of making the engine's power available at low tug speeds as well as high; and it allowed going astern without first stopping the engine. Smit had built a geared diesel tug as early as 1932; the pioneer in Britain was Crosthwaite's two-speed *Caedman Cross* of 1953. Another approach to flexibility was the variable-pitch propeller, controlled from the bridge. Altering the pitch, or angle, of the blades allowed the amount of thrust developed to be varied quickly to suit towing circumstances, giving short bursts of high power for moving a ship, or steady, economical running for the long tow. Alexandra helped to pioneer variable-pitch propellers in Britain, combining them with two-speed gearboxes; then went off them for a while.

It was not only customer demand that made the motor tug. The need for it coincided with cheap oil and dear coal, which made the expense of scrapping one fleet and building another look more realistic. Cheap oil was also a help at a time of shipping recession, when the number of customers was shrinking and economies such as lower running costs were welcome.

There was another new consideration that owed nothing at all to technology. Cities that were also river ports suffered acutely from ships that spewed their acrid black coal smoke into the atmosphere. For a century authority had tried, spasmodically and ineffectually, to curb the steam tug's emissions, which were particularly noisome on account of the quantity of smoke a hard-worked tug produced. In 1956 a Clean Air Act forbade all emission of dark smoke from chimneys, and coal-fired steam tugs in river ports found themselves outlaws overnight. The diesel tug offended far less, and could be worked within the law. Cheap oil and pollution-free fuel favoured the oil-fired steam tug, too. Conversion was expensive - about £10,000 per boat in 1958 - but was a cheaper proposition than building a new tug, so where it could be justified by the general condition and projected lifespan of the boat, steam tugs were made over to oil firing. By this time, however, conversion was at best a stop-gap, for the motor tug had come of age.

Now that power and manoeuvrability were at such a premium, the Kort principle took on a new lease of life; although in an improved form. In 1953, Smit put in service a harbour tug with a Kort rudder. In this the nozzle turned with an integral rudder blade, the whole unit rotating through 45 degrees to either side, enabling the tug to move sideways, and generally easing rapid changes of direction. In reverse, a tug with a Kort rudder could go astern at full speed in a straight line. The Lamey tug *Marie Lamey* of 1956 was the first British tug to be so equipped.

'Empress of Britain' with 'Wellington'.

An alternative for harbour work, Voith-Schneider propulsion, had been evolved as long ago as 1930, by J. M. Voith in Germany, although it first appeared in Britain in 1954 on a Cory tug at Cardiff. In a Voith-Schneider propulsion unit, a vertical shaft leading down out of the hull carries variable-pitch propeller blades mounted vertically about a shaft on a rotating plate, giving it a motion rather like that of an egg-whisk. The blades have variable pitch, to give thrust in any direction, driving and steering the tug at the same time. Boats so equipped were called 'tractors'. To extreme manoeuvrability, the Voith-Schneider system added safety. Used singly, the unit was mounted about one-third of the way back from the bows, thus altering the longitudinal centre of gravity so that the towing hook could be at the stern rather than just abaft the tug's mid-point. This greatly reduced the risk of a tug being girted, or pulled over, when turning a ship. In the mid-1980s the Voith-Schneider system was still unsurpassed for use in dock or river, where the call was for manoeuvrability and short bursts of power. It was not suitable for long hauls at sustained speeds, as at sea, for which the Kort or related systems were more economic, because it exerted less thrust for the same engine output than the Kort rudder, and was more expensive.

The Kort principle was refined to obtain the advantages of Voith-Schneider propulsion. The German Schottel unit was a propeller in an open-ended nozzle turning through 360 degrees so as to form a combined steering and propulsion system impelling the tug in any direction. The Schottel unit was used in pairs, each unit driven by its own engine. It could be fitted at any point on the hull. Mounted one-third or one-quarter of the way back from the bows, it allowed the use of a stern hook, in the same way as Voith-Schneider propulsion. It was introduced to Britain in Crosthwaite's *Greatham Cross* of 1976.

The Japanese Z-Peller system was first seen in Britain on the new Rea tug *Eldergarth* in 1981. It was similar to the Schottel in principle, both using variable-pitch propellers. A third system, the Aquamaster, otherwise almost identical, offered the choice of variable pitch or fixed blades. All three were known generically as azimuthing propulsion systems, from the manner in which the drive train went through three right-angled changes of direction - from horizontal to vertical to horizontal shaft - and then gave drive in any horizontal plane.

The design of the 1987 tug always included one of these five propulsion systems, or a variation upon them, sometimes aided - in the case of sea-going tugs - by thrusters as well. The idea of using thrusters to help tugs seems to have originated in Britain with Ridley Tugs of Newcastle in 1956.

In the mechanics of towing, if not in methods of propulsion, an argument was settled. By the mid-1960s artificial fibre - usually nylon or terylene - had displaced manila in towropes. The argument over their merits had lasted since 1948, when nylon ropes were tried on tugs on the Tyne. Tugmen at first disliked man-made fibre because when it broke, it did so suddenly, without any of the unravelling of manila that gave a few moments' warning. But it was a quarter of the weight of manila, and so easier to handle: a long hawser as used in ocean towing could easily weigh several tons. Also man-made fibre stretched readily. The steel wire of the modern sea towing hawser lacked the elasticity needed to take sudden strains, so a 'spring', or shock absorber, of artificial fibre was incorporated. Finally, weight for weight, man-

A 1930s view of 'Voltaire' and 'Morpeth' leaving dock.

Tug and tow: 'Albert' and 'Queen Elizabeth 2'.

made fibre was around six times as strong as manila; though it could and did part. A nylon spring broke at Port Talbot in 1963, mortally injuring a crewman of the *Cambrian*.

Bulky manila disappeared as towing winches spread from sea-going tugs to harbour tugs. Modern tugs had the choice of a hook and a fixed length of hawser, or the winch that gave the right length for every occasion. More tugs came to incorporate a bridge-operated hook quick-release mechanism, such as the very successful one devised by a superintendent engineer of Alexandra. For lack of a quick-release system, the *Vanquisher* was sunk by its tow in the Thames in 1975.

Technical adaptation to the demands of new shipping could not on its own meet the threat posed by postwar trading patterns. Alexandra's old outports, and their home port of Liverpool, responded in different ways and to different degrees.

Southampton reopened for civilian business in September 1945. Bomb-damaged though it was, it very soon became a national showcase for the return of peace - romantic passenger liners redecorated in prewar colours, cameras welcoming celebrities welcoming cameras. Southampton capitalized on its natural advantages: its proximity to the European ports, its sheltered deep-water harbour, its four tides a day, the freedom of its docks from time-consuming locks. The opening of the brand new Ocean Terminal in 1950 was a sign - happy days were here again. The *Queen Elizabeth* joined the *Queen Mary* on the transatlantic run, as did the ships of the United States Line: the *America* and later the *United States*. Competing with them were the French line's *Liberté* (formerly the German *Europa*), and later the *France*. Until the end of the 1950s, passenger traffic went on expanding. The number of passengers bound for the Far East shrank with Britain's overseas empire and the growth of jet air travel, but the Australian emigrant trade increased; and Southampton was the country's main outlet for trade with South Africa.

More significantly, as it turned out, 1951 saw the opening of another, considerably less glamorous terminal, well away from the tourists and the stars - the Esso Marine Terminal at Fawley, where oil was discharged for a new refinery, Britain's largest. At Hamble a more modest Shell-BP terminal stored and distributed oil products brought in by tankers. By 1967, oil accounted for all but 3.1 million tons of the 28.3 million tons of cargo passing in and out of Southampton. Business for the port's tug operators boomed with it. When Alexandra's tugs returned they found their old partners and rivals Red Funnel battered but intact. With boats lost and new building embargoed during the war, tugs were in short supply. As time went on, more and more were called for. Before tankers began to grow in size, they

Two Alexandra tugs, the coal fired furnaces blowing black smoke over the 'Reina del Pacifico' in the Mersey in the 1930s.

Yet another well-known name. The largest liner to be launched in Britain since the 'Queen Elizabeth', 'Windsor Castle' (1960), 38,000 tons with 'Flying Cock' and 'Pea Cock' ahead. These tugs, owned by North West Tugs, were managed by Liverpool Screw Towing Co.

Departure of H.M.S. 'Ark Royal' from Southampton, 'Flying Kestrel' and 'Brambles' ahead, 'Gladstone' aft, 11th October 1957. Flight deck of U.S.S. 'Forrestal' in the foreground. Note that there are also a couple of paddle tugs in there. (Southern Newspapers Ltd., Southampton).

'Fighting Cock' and 'Game Cock V' both tugs launched on the same day in June 1953 from Cammell Laird. Seen here in 1954 in charge of aircraft carrier 'Ark Royal' also built at the same yard.

needed only one or two tugs each. By the end of the 1950s the number was six.

Most of the growing tanker traffic came Red Funnel's way, for they had the contract with Esso at Fawley. Between 1953 and 1961 they built six new tugs to cope with the business. All needed to be powerful boats, and all had twin screws, which was the Southampton firm's favoured recipe for power and manoeuvrability. In 1953 two oil-fired steam tugs of 1500 hp, the *Hamtun* and the latest *Sir Bevois*, came into service. Two diesel tugs followed in 1956: the 1340 hp *Atherfield* and *Culver*. Pride of the Red Funnel fleet was the tug tender *Gatcombe*, disposing of 1800 hp and 20 tons of bollard pull. Built in 1960, she would carry up to 400 passengers. Four years later Red Funnel put into service their most powerful tug so far, the 1800 hp, 493-ton tender *Calshot*. Two Red Funnel tugs, specially equipped for fire-fighting, were on permanent station at Fawley.

In the early 1960s passenger traffic slowed as more people - particularly businessmen - used the new jet airliner services to cross the Atlantic. As early as January 1961, Alexandra had second thoughts about ordering a new tender for Southampton. That was just as well: the *America* was withdrawn in 1964, although the *United States* remained in service until 1971. The *Liberté* was replaced by the *France* in 1962. In 1967 the *Queen Mary* sailed from Southampton for the last time, followed by the *Queen Elizabeth* two years later. The rearguard of the old giants was the *France*, the longest passenger liner ever built. She, too, was taken out of service in 1974. New ships such as the *Oriana, Canberra*, and *Queen Elizabeth II* were fewer and smaller, and although Southampton was their home port, they spent part of their time cruising instead of shuttling regularly in and out. While the *Queen Elizabeth* had needed seven tugs to move her, the *QEII* usually called for only four.

In 1962 the report of the Parliamentary Committee of Inquiry into the Major Ports of Great Britain, called the Rochdale Report after its chairman, had recommended the development of Southampton's freight trade. The port authorities took this to heart: as the passenger trade declined, freight business was built up. In March 1966 the first container ships of the United States line berthed. Within two years Southampton had opened a terminal and thirty new berths on the Western Docks, all designed for container traffic. The south now had Britain's leading container port. For the tugs, not even the six-week seamen's strike of 1966 was totally disastrous, for they were needed to move ships from berth to berth as docks became congested. Some ships were moored three abreast.

The ups and downs of the South Wales ports were more frequent. Increasingly, Swansea was a port with all its eggs in a few baskets - mainly oil in, and oil products, coal and steel out, with timber and other products a long way behind. It suffered accordingly when trade in these commodities fluctuated. Any port dependent on coal exports had a bad time when shipbuilders went over to diesel propulsion for all but naval vessels and the biggest merchant ships, and to oil-firing for all steamers. By 1960, Britain's merchant fleet consisted of 1833 motor ships and 187 steamships. This was not such bad news for Swansea as for other ports, for it was already on the right side of the fence, with an oil refinery that had been in place since 1920. In 1952, it was oil that sustained a record year for the port's trade, and four years later oil accounted for 80 per cent of its imports.

But although there was a world trend of growth in the oil industry, local hiccups went against trends, with damaging effects on one-trade ports. In 1960 the amount of oil and oil products shipped in and out of Swansea reached a new high, but in 1961 imports plummeted by nearly three-fifths when the new refinery at Milford Haven started pumping direct to the Llandarcy refinery. Then imports through Swansea rose again, especially after Llandarcy expanded its capacity in 1963. Sharp, almost annual rises and falls in business were the story of Swansea's trade throughout the 1950s and 1960s.

Port Talbot seemed to have a more secure future. Reliant as it was on coal exports, Swansea's satellite had led an uncertain existence since the 1950s. Exports had risen from 1.26 million tons in 1951 to 2.3 million in 1955, but by 1959 were down again to 2 million tons. Iron ore was to be the saving of Port Talbot as oil had been at Swansea. The Margam steelworks was extended, and in 1966 building work began on a new deep-water harbour to take the giant ore-carriers that would keep its furnaces fed. Within four years the work was completed. Later the Llanwern steelworks, which had taken its ore from Newport, looked to Port Talbot also. The old dock was closed, and the number of ships calling fell by nine-tenths; but the new work more than made up for the loss. Seventy-five bulk carriers a year were expected, each of about 100,000 tons. In 1961, 2¾ million tons of ore had entered the port. By 1972 the figure was 4.1 million tons.

Liverpool after 1945 never recovered the predominance it had lost between the wars. True, its economy was more broadly based than hitherto, with trading estates manufacturing and assembling for the food, motor vehicle, electrical, and general engineering industries, so it was no longer quite so dependent on its port. This was just as well, for the port was slow to recover from the terrible battering it had taken in 1940-1. The war damage claims of the Mersey Docks and Harbour Board were not met until 1949. Harbour dues, having risen hugely during the war and the immediately postwar period, were actually cut in 1951, and did not rise again until 1953. In consequence, there was not enough revenue to complete rebuilding until the 1960s.

The tonnage of peacetime shipping using the port did not attain its 1939 level again until 1952. In 1949, Merseyside handled 28.5 per cent of the nation's trade, compared with London's 31.5 per cent. At this time it was

At a time of expansion in 1950, three new steam tugs were ordered from Cochrane & Sons, 'Canada', 'Gladstone' and 'Formby' seen above, all 1,200 h.p. and 237 gross tons (Studio City, Hull).

'Corinthia' and 'Nelson' (Elsams Ltd., Liverpool).

still Britain's top exporting outlet; but in the mid-1950s London overtook it in this respect as well. Frequent and damaging labour disputes, as well as restrictive goods handling practices, did nothing to encourage shipowners or their customers to use Liverpool. The consequence of slow recovery, hampered by labour strife and labour costs - which had multiplied six times in twenty years - was visible in a further drop in Liverpool's share of British trade. By 1965 it was only 18.3 per cent, and the port handled barely a fifth of exports. Increasingly, too, Liverpool suffered from its geographical situation, far from the markets of the rich south-east and continental Europe. So long as there were ports there to handle them, it was more economic for bulk cargoes and containers to discharge nearer their ultimate destination.

Growth was there, but it was slower than at other ports. The tonnage of ships in and out of the port rose from 21.1 million in 1950 to 29.3 million ten years later, and its earnings between 1945 and 1965 doubled. Imports of oil were largely responsible, benefiting a few ships and a few new docks. The Queen Elizabeth II oil dock, for tankers of up to 30,000 tons, was opened in 1954; but the size of tankers increased so rapidly that it was soon unable to take the biggest. An oil refinery and jetty at Tranmere, next door to Birkenhead, followed in 1960. It would take 100,000-ton tankers, and eight years later was developed to accommodate vessels twice that size. Oil imports climbed from half a million tons a year to 6.3 million. By 1964 the figure was 7.9 million tons, or 48 per cent of all imports.

In 1967 new berths for container ships were opened in the Gladstone Dock. Five years later a massive container terminal was opened at Seaforth, on the seaward side of the Gladstone Dock; and another container terminal followed upriver at the independent port of Garston in 1979. But the Mersey was not as susceptible, physically or geographically, to development as some other ports. In 1962 the Rochdale Report had already recommended that Liverpool should not be developed further, and being written off at such a high level did not help when the Mersey's troubles deepened. Meanwhile other ports with lower costs and more favourable locations gained ground at Liverpool's expense.

The vicissitudes of the port were mirrored in the ups and downs of its tugowners. When the wartime convoys sailed away for the last time, revenue fell while costs soared as long-delayed repairs were put in hand. When at last the port was rebuilt and trade reached peacetime levels again, the tug operators tried to rebuild too. Well managed companies such as that of Alexandra did well into the second half of the 1960s; all added to their fleets, and modernized them to attract the big new ships.

Modernization was long overdue. In the mid-1950s, at the beginning of the growth in ship size, seven-eighths of Britain's tug fleet was still of prewar vintage. Because it was so old, little was set aside in company accounts for depreciation, making costs appear artificially low, and profits misleadingly high, which was dangerous at a time when a large-scale, high-cost replacement programme was imminent.

Between 1951 and 1956, J. H. Lamey bought four old steam tugs. Because they were expanding into coastal towage, they equipped one of the boats, the *B. C. Lamey*, with radar which was useful, too, when working in a crowded estuary in darkness or fog. They converted two of their acquisitions to diesel propulsion in 1957. One, the *John Lamey*, was the first motor ship tug on the Mersey. A year earlier another of these purchases, the *Marie Lamey*, had been given the first Kort rudder seen on a British tug. Lamey were so confident that in 1959 they built their very first new boat, the *William Lamey*. She was a 166-ton motor tug with direct engine control from the bridge, and the novel feature of twin slim exhaust uptakes, instead of the now redundant single large funnel, affording an excellent rearward view from the bridge. Four more new tugs followed, the last in 1968; by which time there were eleven boats in the fleet.

The ownership of the Liverpool Screw Towing and Lighterage Company underwent a change in 1951. It had been the manager of its tugs, which were still owned in sixty-fourth shares by individuals, and also by prominent shipping lines such as Bibby, Clan Line, Ellerman, and Brocklebank. These institutional shareholders, concerned at a lack of investment in new tugs, now set up a new company, North West Tugs, to take over completely the ownership of the fleet. Its major shareholder was Cammell Laird, builders of more than half the Cock fleet and providers of much of its business. Cammell Laird were still busy at Birkenhead, launching passenger liners and warships such as the aircraft carrier *Ark Royal*. The Liverpool Screw Towing and Lighterage Company continued to manage the tugs; but the customers, like others before them, had decided that it made better economic sense to possess their own. Like others before them, they were mistaken.

By 1956 customers were refusing to use some of the older boats. 'Morpeth' built in 1901 was one of those consigned to the scrapyard.

'Jalasiltonhall' with a head line to 'North Light'.

Under the bow of 'Landwade' is 'North Buoy' one of the last two steam tugs ordered by Alexandra from Scott of Bowling in 1958.

Two new steam tugs were launched at Cammell Laird in 1953 - the *Fighting Cock*, and the 1250 hp *Game Cock V*. At this time the fleet numbered thirteen boats. The company's barges were sold in 1956, and its name shortened accordingly. The first purpose-built motor ship tugs on the river, the *Heath Cock* and *West Cock*, entered service in 1958. Between late 1959 and summer 1960 three 1400 hp diesel tugs followed. One was the *Weather Cock*, which was the first Mersey tug with two-speed gears.

On a smaller scale, in 1958 Furness Withy's subsidiary company Johnston Warren built themselves three new, powerful tugs to tow the ships of their allied line, Shaw Savill - work that Alexandra lost as a consequence.

William Cory & Son, in the guise of the Rea Towing Company, were the first Mersey operators to build a new tug after the war: the *Aysgarth* of 1950. Six new boats were delivered between that year and 1954 - all coal-fired steam tugs, converted to oil firing between 1956 and 1959. In 1958 the company had set their first motor tug to work, the *Hazelgarth*. In 1960, Rea won a contract to tow Shell tankers at Tranmere; and in 1965, Shell went back to them for tugs. Nine years later Rea put into service at Tranmere Britain's two most powerful diesel harbour tugs, each of 4,000 hp and 45 tons bollard pull, to tow Shell's 200,000-ton VLCCs.

Cory seized opportunities outside Liverpool too. Milford Haven, one of Britain's finest natural harbours, had been little exploited commercially. Before the war it had been best known for its imports of fish, and handled less than half a million tons of shipping a year. It had begun to take in oil imports even then; but its oil boom began in 1960 when British Petroleum opened a new tanker terminal, and Esso a refinery. Three years earlier R. & J. H. Rea Ltd, a member of the Cory group based on the Bristol Channel ports, had offered four purpose-built firefighting tugs to move the tanker traffic, and got the contract. By 1962 they were in service. In 1963, the company bought the tugs of the defunct Bristol Channel Towage & Salvage Company, and also took over the British Transport Dock Board tugs at Cardiff and Barry, the service run in pre-nationalization days by the Great Western Railway. In 1969, Rea were awarded the contract to supply four specially-designed firefighting tugs to handle 300,000-ton tankers at Gulf Oil's new terminal at Bantry Bay, on the other side of the Irish Sea.

In the first months of peace, as the wartime shipping boom ended, Alexandra's weekly revenue dropped by more than 60 per cent, to well below its 1939 level. By the end of 1945 it had recovered to around £5000, and thereafter climbed slowly but more or less consistently. When the company's Southampton office reopened for business, it started to bring in around a quarter of the company's income, with the Swansea operation contributing about the same amount, but Alexandra's business here was patchy. In July 1946 two Alexandra tugs and all three Britannia boats were laid up for a while. Tug business at both ports was static, while at Liverpool it grew. Out of a total weekly revenue of around £8000 in 1949, more than £5000 came from the Mersey. Port Talbot brought in a few hundreds.

On the other hand, the wages bill went up by nearly 40 per cent, and the fuel bill by more. In early 1949 coal cost up to 60s a ton. Labour troubles became endemic to all ports. More often than not, union demands were for more time off - which was less easy to arrange than wage increases in an industry whose working hours were governed by ship movements and tides. At Southampton the tugmen twice ignored settlements reached between employers and union, or rejected the outcome of supposedly binding arbitration.

Repairs to a war-worn fleet, essential if customers were to be gained or retained, totalled £76,000 in 1946 - more than £1400 a tug on average. The essential restoration was done, so maintenance costs fell back for a couple of years before increasing again; but repairs were a persistently heavy item of expenditure as the fleet grew older, and was worked harder. To keep its head above water, Alexandra had to increase towage rates twice at Liverpool and once at Swansea and Southampton. In these years, annual revenue went up by a little more than 40 per cent. Taken with the wage and fuel bill, this figure suggested both an accurate assessment of the proper rate levels, and how little leeway there was. There was just one morsel of comfort. The shipowners, basking in a sellers' market and rising freight rates, tended not to argue as tenaciously about towage or salvage charges as hitherto. Indeed, when three Alexandra tugs at Southampton saved the *Queen Elizabeth* from stranding in 1947, Cunard made up the company's expenses claim of £2886 to £5000. Alexandra passed the balance on to the crews.

Needless to say retrenchment, not expansion, was the theme of these years. No new tugs were built. In 1948 the *Flying Kestrel* had to go for scrap, calling for a replacement, but it was a second-hand ex-Admiralty tug that became the 'new' *Flying Kestrel*. Rationalization was in the air already, particularly in volatile Swansea. Here Alexandra made their first acquisition since Jolliffe nearly forty years before. By now the Dewsburys had only one tug left: the *Benson*. The others, on charter to the Great Western Railway or the Admiralty, had left the port, never to return. In 1947, Alexandra bought out the Dewsburys, and the *Benson* passed back into the Alexandra fold. In August 1949, Alexandra came to a work-sharing agreement with Britannia, as they had done with Rea and Red Funnel.

Restless men and ageing tugs, worked hard, suffered accidents. In December 1945 the *William Poulsom* was sunk at Southampton with the loss of four lives when she was overrun by the Royal Mail liner *Highland Monarch*. The following summer another ship sent the *Coburg* to the bottom of the Mersey; while in January 1948 the *Toxteth* went down, taking with her all but one of her crew.

In 1964 Alexandra went to a builder new to them, Richard Dunston of Hessle — for three new boats. One of these was 'Brockenhurst' seen here with the old 'Foudroyant'. (Armitage (Photographers) Ltd. Southampton).

Quayside view of 'North Beach', 'North Wall' and 'Nelson' overshadowed by the 'Empress of Canada' at Princes Stage, Liverpool. (F. Leonard Jackson, Heswall).

An unchanged team of directors presided over this difficult time. Early in 1947, Frank Stark, who had founded and built up the Southampton end of the business, died after more than half a century with Alexandra. In December of the following year a minor revolution occurred in the Liverpool office, where Henry Brussel Bicket, son of Alex, after serving six years as a rating and sea officer in the Royal Navy, went to work in the accounts department under William Henry Paynter, the cashier. For the first time, a future director who was a member of one of the ruling families was being put through the business from the bottom, instead of arriving on the board from outside, fully fledged.

For Liverpool, for its tugowners in general, and for Alexandra in particular, 1950 heralded better times. The increase in revenue, which had been slow, rose in the year by 9 per cent. More remarkably, the wage bill actually went down a little, the fuel bill rose by less than 3 per cent, and the repair bill fell by £35,000. This was just the beginning of a boom for Alexandra that far outstripped the modest rise in the fortunes of the port. By 1955 their income had trebled in ten years, which said something for their management at a period when the trade of the port was growing only a third as fast. Annual revenue passed the £1 million mark in 1957; for 1958-9, the net profit was almost eight times the 1949-50 figure. In 1949, income net of the three major items of expenditure had been £70,000, or 17 per cent of gross income. By 1959, at a time of negligible general inflation, the corresponding figures were £479,000 and 44 per cent - in spite of the fact that the wage and fuel bills had more than doubled in the same period. Profits and reserves paid for three increases in Alexandra's share capital, which were made by issuing bonus shares to shareholders. In 1951 the capital was doubled from £50,000 to £100,000; in 1957 it was increased to £400,000; and in 1959 to £600,000.

The explanations for this growth rate are not hard to find. Part of it lay in the outports, where the increasing prosperity of the 1950s was having its effect. Towage rates could be increased, with rising costs (especially wages) as justification. In 1950 the Southampton rates went up by a third, and those at Swansea by a quarter. Thereafter rises were less dramatic, but by 1956 were biennial and sometimes annual events at all ports. Revenue grew accordingly. At Swansea, as in Liverpool, it nearly doubled over the ten years; while at Southampton and Port Talbot it almost trebled. In the second half of the 1950s, salvage grew as well.

One sign of good times was the institution of the '24-hour tug', kept available for work around the clock. Manning varied slightly from port to port, but basically each 24-hour tug had two crews, each of which, on weekdays, was alternately on duty and off for twenty-four hours. At weekends, one crew was available while the other was off duty. The system avoided the necessity, which might otherwise arise, of a tug crew working seven days a week at busy times; while for the owners, it was cheaper to pay for more crews than for more tugs. But, of course, the system depended on a strong, steady flow of work around the clock. Southampton never had it, because passenger liners came together and went together, with troughs in between. Crews who were not on twenty-four-shifts had relief crews to allow them nights off duty. There would normally be standby and spare tugs too, to cater for sudden demand, and to allow other boats to be taken out of service for repair. In Liverpool by October 1965, there were fifteen 24-hour tugs and two relief tugs in the Alexandra fleet.

It was time to expand. In late summer 1950 three new steam tugs were ordered from Cochrane & Sons of Selby; the first since 1936. They were the latest *Canada, Formby,* and *Gladstone,* delivered in 1951. In addition, an Admiralty tug, sister ship to the latest *Flying Kestrel,* was brought to Southampton and renamed *Brambles*. Meanwhile a landmark passed from the scene when the old *Canada* was sold for breaking, after fifty-five years of service: the last nineteenth-century relic of the company, apart from a few employees. Nowadays, an Alexandra tug did not have to be sold to enable her name to be used for a new boat: the company renamed her when she was moved to a new port of registration, freeing the old name for a new tug. For instance, the old *Formby* and *Gladstone* were renamed *Gower* and *Langland* - local Welsh placenames, like *Brambles* - when they were moved to Swansea.

Eighteen months later, another three tugs were ordered, perhaps because of an increase in work at Port Talbot and Southampton. They were the *Waterloo, Wallasey,* and *Canning*. The last-named was to be oil-fired. Cochrane had been asked for prices for diesel tugs, but had quoted £104,000, against £70,200 for steam. Alexandra, who found both figures 'staggering', went for steam. In September 1954 a further pair of tugs was put in hand, by George Brown of Greenock, at a price of £87,600 each. There was no shock this time: perhaps Alexandra had accepted the inevitable. These tugs, the *North Light* and *North Rock,* introduced the more powerful 'North' class of boat, of 1050 hp and a bollard pull of 13½ tons. Only three months later, three more were ordered: two, the *North Beach* and *North Quay* from Scott & Sons of Bowling on the Clyde, and one, the *North End* from W. J. Yarwood & Sons of nearby Northwich in Cheshire - all for delivery in 1956. Diesel tugs were still 'of no interest to the company'.

But under the press of events, this attitude was changing. In July 1956 motor tugs were at least of sufficient interest to be 'discussed' by the board. Another pair of steam tugs - Alexandra's last - was ordered from Scott in 1956: the *North Buoy* and *North Wall*. Construction of these and the previous pair was accelerated because of pressure from customers, who were refusing to use the older boats. These were being got rid of: the *Canning, Morpeth,* and *Mumbles* (the former *Waterloo*) went to the scrapyard when new boats came into service. In 1952 the *Brunswick* was sunk by the Ellerman liner *Ionian*

'Camel No. 7' being looked after by three Alexandra tugs sporting different style funnels. (Elsams Ltd., Liverpool).

Fine example of the powerful 'North' Class. 'North Beach' was built by Scott and Sons of Bowling in 1956. Her bollard pull was 13.5 tons, miniscule by today's standards. (A.T. Kelly, Glasgow).

off the dock she was named after, drowning two of her crew. She was raised, then sold. In spring 1957 the *Collingwood*, also rammed and sunk off the Brunswick dock, was abandoned when it was found that she would cost too much to repair.

In the same year, Alexandra took the plunge. The two new tugs ordered from Yarwood for delivery in 1958-9, the *North Isle* and *North Loch*, had diesel propulsion with bridge control. In addition, three second-hand boats were bought from Furness Withy's Johnston Warren line for £55,000, to replace still older tugs at Swansea. They were renamed *Mumbles, Margam*, and *Murton*. The *Victoria* and *Brocklebank* went to the breakers' yard soon after, followed in 1959 by the *Cambria* (ex-*Wallasey*) *Trafalgar, Gower* and *Harrington*. The *Harrington* had been the death of two of her crew when she sank while towing a tanker in the King's Dock at Swansea in 1954. Since 1951, the fleet had been transformed. More than a third of it had been scrapped, and fifteen new boats delivered. With the four second-hand acquisitions, there had been a net gain of seven, so it was a stronger fleet in numbers as well as quality. At the same time, a revolution was getting under way: the first motor tugs had joined. The *Romsey* had been converted to oil firing in 1956, and now work began on converting three more tugs: the *Canada, Formby*, and *Gladstone*.

Confidence was total. The traditional kind of business was growing; and as for the new ships in Alexandra's ports, it did not seem that they so differed from the old that a gradual development of the existing tug type would not be able to cope. Portents were ignored; and why not, when there was more than enough business of the usual sort? In 1955 the Milford Docks Company asked Alexandra for two tugs to be stationed at Milford Haven. They were turned down: tugs would be sent from Swansea as and when local work allowed. As a result, others got in. By 1956 a tug of the Overseas Towage & Salvage Company was based at Milford Haven, but Alexandra decided not to compete. As a result, Rea eventually took over towage in the port. And in 1957, when Esso came shopping for 1500 hp tugs for a new oil refinery under construction at Cork, Alexandra did not bid. Their existing tugs were too small, and they had too much work at their own ports.

There, it was a story of victory over heavy odds. New tugs now took sixteen months or more to deliver - twice as long as prewar - owing to many of the same troubles that afflicted all British shipyards, such as higher wages and dearer materials. This meant that the economies and the higher revenue that they brought were delayed, and that their cost rocketed. A new steam tug in 1950 cost £59,000; around four times the 1936 price. Boats ordered for £70,200 early in 1952 cost over 8 per cent more by the time they were delivered, even after the objections of the customer had been met by reductions. Both parties came to realize that the yards could not stick to contract prices, and adopted saving clauses in their agreements.

Prosperity brought problems. Revenue - and more important, the confidence of shippers - was hit by the growing number of lightning tug strikes at all ports. All the Mersey tug crews - 630 men - came out for a week in October 1959, sixty ships were held up, and Alexandra's income from the river sank to little over £1000, from the usual £12,500 or so. Most of the disputes now concerned money, and usually a rise was conceded. More often, the threat of action did the trick: there were more than twice as many increases as actual strikes, half of them at Liverpool. Dock strikes and seamen's strikes had much the same effect on operators: by stopping ship movements, they stopped the flow of revenue as effectively as a walk-out by tugmen. During the Liverpool dock strike of October 1954, the company's Merseyside income fell by more than a third.

A basic forty-eight-hour week at Liverpool and an expensive company pension scheme for all 'floating' staff - extended soon to shore staff - were conceded in return for an undertaking not to press for higher overtime rates. By 1957 all the ports were working a basic forty-eight-hour week.

Fuel prices went on climbing. Coal reached 64s a ton in 1951, and 81s two years later. Even the price of oil, which became significant as oil-firing spread through the fleet, continued to rise until the mid-1950s, reaching £11 19s a ton early in 1957. Then came 1958, when the price of oil fell three times in the year. The age of coal firing was finally over.

The rise in Alexandra's growth curve flattened, and by 1968 a crisis point had been reached. Revenue went into steep decline: income over 1967 showed a drop of nearly 15 per cent. Yet, just as income became erratic, money had to be found to ensure long-term economies - money for tug building, for acquiring other tug fleets, and for rationalizing operations. It became progressively harder to find from the company's own resources. The shipping recession began to bite early in 1963. Over the next eighteen months there was a modest overall increase in revenue, but it did not, in percentage terms, match rises in towing charges, which customers were once again resisting. And in 1966 corporation tax - a new, permanent profits tax - was introduced.

Expansion was hampered because costs gobbled up revenue. At Alexandra there was a sense of drama - of essential growth struggling to keep ahead of the financial consequences of the industry's increasing difficulties. Although annual revenue rose every year until 1968, with the exception of 1960-1, this did not give a true picture on its own. Revenue net of wages and fuel went up only 1 per cent between 1960 and 1965. Repairs remained a sobering item, in spite of new tugs - the bill in 1964 alone was almost £1 million. If this item were included, the equation would have looked sicker still. Towage rates went on rising, but usually did so in response to loss of business, wage increases, and cuts in basic working hours: forty a week at all ports by 1968. They did little more than keep revenue high enough to hold the ground won.

'De Grasse', a French cruiser, approaching Princes Stage, Liverpool with 'Trafalgar' in attendance 13th June 1969.

Alexandra's Liverpool operation was locked more firmly in the spiral than any other. There were cuts in hours and wage rises for the tugmen in 1964 and 1966. There was a ship repairers' strike in 1961, the week-long dock strike of 1965, and the six-week seamen's strike in summer 1966, when those ships that still moved were blacked by the tugmen. More destructive of trade and confidence was the six-week dock strike of 1967 - brought on by the dockers' dissatisfaction with the terms of their 'decasualization'. To make up for the loss of revenue that all these entailed, towage rates went up every year between 1963 and 1968. The effect of the shipping recession was especially sharp on the Mersey. By the end of 1963, there was a tug laid up for the first time in many years, and crews laid off. Another tug was mothballed early in 1964, and that summer three more crews were made redundant. Their union accepted the cuts, but the Liverpool crews did not. In autumn 1965 some began to look away from the T&GWU towards the dockers' breakaway union, The National Amalgamated Stevedores & Dockers. It was the early 1900s all over again.

Southampton, too, preserved its reputation for militancy, extracting rises or cuts in hours in five of the eight years 1960-7. The Swansea tugmen won annual rises in the years 1964-8. There were strikes at Port Talbot in 1964-5, not involving tugmen directly but stopping their work, so that income was drastically reduced. The tugmen themselves struck comparatively seldom, but they came out at hard-pressed Liverpool in February 1967. The financial result was 'grievous' - not in terms of business lost, for it was brief, but in the concessions Alexandra were forced to make to end it quickly. By now, wages absorbed 40 per cent of the company's revenue.

The main growth, contributing to the rise in revenue and saving the figures from looking much worse, came not at Liverpool, where income was static in spite of frequent rate rises, but at the outports. There rate rises were fewer, and revenue generally healthier, even though growth was erratic. In the early 1960s Southampton was Alexandra's 'flagship' port. It was allotted the newest tugs, and new offices were built for the company; though with long delays it was 1965 before Alexandra moved in. By that time the passenger liners were vanishing fast, but business was good for the company in 1966, thanks to growing tanker traffic. At the Swansea office revenue tended to follow the dizzying ups and downs of the port's trade. In 1961 income took a bad knock when oil imports fell; but in 1966, Alexandra's good year at Southampton, their Swansea income rose too, with increased output at the Llandarcy refinery. In 1968 revenue went on improving at Southampton, that least depressed of Alexandra's ports; there, the company won a towing contract from Elders & Fyffes. But in Swansea in July, a tug was taken off and two crews made redundant because business had slumped.

There were some bright spots in the gloom. Alexandra won a major

'Egèrton' and 'Langton', seen above, with their sister-ship 'Brocklebank' were a new type built by W.J. Yarwood in 1964. (L.A. Baker, Birkenhead).

This gives some idea of what it feels like on the aft deck of a tug, ready to tow the ship 'off the wall'. The gog rope can be seen hanging over the main towing spring.

contract to transport 2 million tons of stone from the Dinmor quarries on Anglesey to the Mersey, for the construction of the Royal Seaforth container terminal. Four tugs were needed full-time for a year, making 2000 trips, each with a 160 ft barge loaded with up to 1300 tons of stone. At around £250 per tug per day, this caused revenue and profits to climb from September 1968; but only while the contract ran.

As its disposable income declined, the company depended more and more for its expansion on its bank. If public policy allowed, the bank would always lend. It required punctual repayments of overdrafts, but would increase them in a good cause, such as the purchase of new tugs or old companies, a reflection of good management on the part of the borrower. Another providential source of the cash tugowners could not find from its own coffers was the government loan on favourable terms, in the shape of the Ministry of Transport's Shipbuilding Credit Scheme, designed to help Britain's ailing shipyards by making it easier for shipowners to build new. The interest rate was less than 5 per cent. Other government aid came from the Board of Trade Investment Grants Scheme, which contributed substantially towards the cost of new tugs. There was good news, too, from South Wales. Alexandra was promised all towage of the bulk carriers at the new Port Talbot deep-water harbour once it was completed. Four tugs would be needed on station, instead of two as hitherto.

Best of all, in the long term, seemed the fact that - with occasional hiccups such as that brought about by the closure of the Suez Canal in 1967 - the price of oil went on falling. In 1961 a comparative study revealed that while a coal-fired steam tug cost £5644 to feed for a year, an oil-fired tug cost £2101, and a diesel tug just £919. If any inducement were needed to press ahead with conversion to oil firing and the construction of motor tugs, this was it. By January 1963 fuel oil for steam tugs was £6 2s 6d a ton, and diesel oil £5 18s a ton - a fall of more than half since the mid-1950s.

With the onset of more difficult times in the 1960s, Alexandra's search for economies led them to seek rationalization at the ports - unnecessary in the prosperous fifties. Little Britannia at Swansea had not done so well even then. In 1954, the Clan line, dissatisfied with their service, had gone over to Alexandra. Britannia could ill afford to lose the Clan custom, for most of the big Swansea lines - Royal Mail, Blue Funnel, Holders, Ellerman - also favoured Alexandra. In 1961, galvanized by a Cory approach to Britannia, Alexandra began negotiations for a takeover which came to fruition in November, when Alexandra bought out the Britannia Steam Towing Company for £52,500. The three Britannia tugs, *Clyneforth, Graigforth,* and *Brynforth,* joined the Alexandra fleet, making ten tugs in all at Swansea. Alexandra now had a monopoly of tug work in the port, and could benefit from economies of scale.

At the same time Alexandra tried to reanimate the stagnant Liverpool tug trade. Passenger liners no longer called, so the *Flying Breeze*, the last tender on the Mersey, was scrapped, ending a story that had begun in the days of sail. It was not just that business was static and rationalization was needed: the company's rivals were gradually increasing their share of what was left. By the 1960s Alexandra's share of traffic in the Mersey work-sharing agreement had fallen to 43 per cent, Rea was in second place with 29 per cent, and Cock Tugs had 28 per cent. In May 1962, Alexandra bought one-third of the shares in J. H. Lamey Ltd for £50,700. In 1963 Alexandra bid for the three Furness Withy tugs at Liverpool, owned by the Johnston Warren line, but failed. This was unfortunate: in May, Royal Mail and the PSNC pledged their business to Furness Withy. Four years later North West Tugs, the Liverpool Screw Towing Company, and the Cock tugs were acquired for £655,000. North West tugs was forgotten, but the older names were kept on for their goodwill value.

W. H. Lamey were not party to the Liverpool work-sharing agreement; in the tradition of their ruling family, they were winning business away from the other Mersey operators as busily as ever. This caused particular anguish at Alexandra, who, since making their new investment, were now in a sense competing with themselves. Goaded by Lamey's 'persistent poaching', Alexandra acquired the remaining two-thirds of Lamey's shares for £240,800 in April 1968, and Lamey's eleven tugs passed under Alexandra control. There were now talks about 'integration' schemes with Rea, and also with Red Funnel at Southampton; these did not develop, but as more and more Mersey ship tugs came under their wing, Alexandra could begin to produce the economies that the situation demanded. By the end of 1968 four Alexandra tugs, three Cock tugs, and one Lamey tug had been laid up or sold, and ten crews made redundant.

By 1960, Rea were established at Tranmere. At Milford Haven, Alexandra found that they had missed an opportunity: the Overseas Towage & Salvage tug was still there, and the Rea tugs had moved in beside her. At Bantry Bay, Alexandra competed for the Gulf Oil contract, but Rea won it. The Rea group's enterprise in seeking out specialized business with purpose-built tugs had paid off; a lesson that Alexandra absorbed for the future.

Inevitably, increased costs demanded more revenue to compensate. Yet this would not be forthcoming from rate rises unless ship owners were offered modern tugs. The modernization of the fleet pressed ahead. Two new diesel tugs were ordered from Yarwood in 1960 - the *Talbot* and *Gower*, each with two-speed gears, a Kort rudder with variable-pitch propeller, and engine control from the bridge. They cost just short of £100,000 each. Another new tug building at Charles Hill's Bristol yard was bought, named *Cambrian*, and sent to Port Talbot. Scott of Bowling were given the job of converting a British Petroleum fire tug into a tender for Southampton. She was rechristened with the famous old name *Flying Breeze*, now that it was

'Herculaneum', another Yarwood built vessel. Her bridge control gave constant trouble, which made Alexandra think in terms of more conventional, simple ships. (Elsams Ltd., Liverpool).

View from the River Weaver of 'North Loch' ready for launching from Yarwood's in February 1959.

again available. This was an ill-judged move, for the conversion cost £100,000, and the boat did not enter service until July 1962, by which time the writing was on the wall for Southampton tender work. She was withdrawn in 1967.

Three more tugs were converted to oil firing. By the time a new *Herculaneum* was put in hand at Yarwood's yard in 1961, the price of a motor tug had gone up to £113,000. She and her successor, the *Alexandra* (ordered the following year), had the same features as the *Talbot* and *Gower*, but were more powerful. In 1962 an intensive programme of tug building was decided upon - fourteen new boats to be built in seven years, and all the oldest ones replaced. Design was to be changed radically. All the new tugs were to be of simple, straightforward, standardized construction, to keep their cost down and avoid the need for special training when crews changed tugs. Furthermore the *Herculaneum*'s bridge control had given constant trouble; and it was found that Kort nozzles became easily fouled with foreign bodies. Delivery times had in most cases improved; but cost remained an intractable problem. In 1961, when the price of a new tug had almost doubled in eleven years, Alexandra had gone shopping for one tug in Holland, for the first time. For the present, however, their fleet remained British-built. The programme was launched in 1963, with orders for six boats with direct drive, conventional propulsion, fixed propeller blades, and control via engine room telegraph in the time-honoured manner. Although they were slightly more powerful than their predecessors, with 1200 hp and 17 tons of bollard pull, the re-design was widely regarded as a retrograde step.

Three of the new type came from Yarwood - *Langton, Brocklebank,* and *Egerton*; and three from a constructor new to Alexandra, Richard Dunston of Hessle on the Humber - *Romsey, Brockenhurst,* and *Ventnor*. The last was delivered in 1965. By now, the building and conversion programmes were paying off: the savings in both fuel and repair bills were noticeable. Also in 1965, a small ex-Admiralty motor tug, the *Sunnyside,* was bought and renamed *Trover,* and another pair of tugs was put in hand at Hessle, to the same specification: the *Trafalgar* and *Nelson,* delivered in 1966. With 1430 hp and bollard pull of 21 tons, they were, however, much more powerful.

Much more would be demanded of Alexandra for moving the great ore carriers at Port Talbot, so in 1967 the new *Mumbles* was ordered from Dunston. She was a departure for Alexandra, in another class from their ordinary harbour tugs. With her 2190 hp and 33-ton bollard pull, she was far and away more powerful than any of her predecessors. She would carry a crew of ten, and set a record for price, too - £216,000. Second-hand tugs were also added to the Alexandra fleet in these years. An oil burner was bought from the Mersey Docks & Harbour Board and renamed *Caswell*; a more interesting acquisition was the ex-Crosthwaite *Caedmon Cross,* rechristened *Margam*. She was soon sold, for the company had gone off geared tugs.

The object was to create a modern fleet, out of a mixed bag that by 1961 still had an average age of twenty-six years. Between 1960 and 1968 twenty-four Alexandra tugs were scrapped or sold, as were all three Britannia boats, an unprecedented slaughter that completed the transition from coal to oil. By this time the combined Alexandra, Cock, and Lamey fleets on the Mersey totalled thirty-three tugs, fourteen of them Alexandra. The number of tugs on the Mersey was little more than half what it had been in the 1920s. Only Alexandra and Rea were left, for at the end of 1968, Rea bought the Johnston Warren/Furness Withy tugs. There were six Alexandra boats at Southampton, and eleven at Swansea.

The cutback at Liverpool was none too soon, or too great. By 1968 there was increasing call for more powerful - and fewer - tugs, for example from Shell at Tranmere. In December that year it was estimated that the container ships of the United States line, now arriving at the Gladstone Dock, would each need no less than five of the existing type of Alexandra tug to handle her.

Presiding over this time of crisis was a reconstituted team. In January 1960 three new directors were elected to the board: Alexander Paul Lovel Mack, known as Paul, a nephew of Charles; the thirty-seven-year old Henry Bicket; and Harvey Graham Mack, known as Graham, thirty-six years of age, a nephew of Harold, and a practising chartered accountant. The last two had been directors of United Grain Elevators since 1951. Henry Bicket was always the practical tugman, who managed the men and boats and found work for them; Graham Mack managed the company's money. Both did so skilfully, which was fortunate, because things were going to get worse before they got better. Alex Bicket began to shed responsibilities. Harold Mack became deputy chairman in June 1960. In July 1963, Henry Bicket and Harold Mack were made joint managing directors. In July 1967, Alex Bicket retired from the chair, at the age of eighty-two; though he remained on the board. Harold Mack replaced him as chairman. Two years later Alex Bicket resigned from the board. In January 1968, Charles Mack died.

The directors new to the board in 1960 had brought new initiatives, and as the days of coal as an economical fuel came to an end, hastened by the Clean Air campaign, new means had been sought to modernise the fleet. Conversion of coal fuelled tugs to oil was the first move, the next was to organise the company's finances on a long term basis. A ten year cash flow projection had been submitted to the company's bankers, incorporating the new low-cost finance offered by the Government to boost British shipyards, and a substantial rebuilding programme set in motion. W. J. Yarwood and Richard Dunston had been commissioned to build tugs, sometimes two at a time, for many years to come. Thus with a combination of new building and company acquisitions, and in spite of a falling real demand for tugs, the company was able to steer into the 1970s with success. Rises in rates

Exciting view of 'Brocklebank' being launched form Yarwood's stocks into the River Weaver. A heart-searching moment for all concerned.

to shipowners, at first merely catching up with wage awards, gave a lift to operating profit margins. A consistent policy was developed whereby tugs were either operated at a profit, or sold and the crews made redundant. The greatest achievement was to be the purchase in 1974 of London Tugs, where there was considerable scope for rate increases and reorganisation.

12

Sun Anglia

Diverse Developments

The Port of London had been experiencing the same crisis as Liverpool, and for many of the same reasons, if on a less dramatic scale. This port was a limb of its city, not its heart.

Wartime destruction was massive, but less ruinous and more quickly repaired than on the Mersey, where the docks were so much more concentrated and vulnerable. And London had other geographical advantages over Liverpool. With economic recovery in the 1950s and 1960s, it had space to grow and deep water, downriver at Tilbury and the oil terminals beyond. This was just as well. The big bulk carriers and container ships could not reach upstream to the heart of the old dockland, which decayed. The Rochdale Report recommended major transplant surgery: the upstream docks should be closed, and Tilbury developed to create a modern port with modern cargo handling technology. No doubt it would have happened even without the report, which merely gave official blessing to a shift in the river's centre of gravity that had begun in the 1920s.

The process was gradual, as in other ports. The close-down of the upstream docks did not begin until the mid-1960s, and did not finish until the end of 1985, when the 'Royals' finally shut down - the King George V, Royal Victoria, and Albert Docks. In the interim all the others had gone - St Katharine's, London, India, Milwall, Surrey Commercial, and the rest.

Between the war's end and the 1960s the traditional types of ship, by and large, frequented the Thames. The big oil tankers had never come upstream, and the cargo liners, tramps, and passenger liners of prewar days were still in the ascendant. This was a good time for the port of London, as for most others. In the grim year of 1948, 41 million tons of shipping sailed in and out. By 1954 the figure was 69 million tons, and in 1961, 89 million tons. By that time most of the growing ship tonnage was accounted for by ever larger tankers. In the same year more than 50 million tons of cargo passed through the port; but it was not a figure that matched the best of the prewar years, and it was never exceeded.

The second Rochdale recommendation began to take effect in 1968, when the first berth of a new container terminal was opened at Tilbury. Small container ships as well as large were drawn there. The owners of these 'feeder' ships, that distributed the containers brought into the continental ports by their bigger sisters, were pleased to unload at Tilbury rather than make their way further upstream, for it was only seven hours' steaming from Rotterdam to the Sunk light.

All the docks, not just those upriver, suffered from endemic labour strife. Industrial action such as the eight-week London dock strike of 1967, and restrictive practices designed to exclude modern freight handling techniques, meant delays that drove ships to Rotterdam or Antwerp. Even where a dock was not too small or shallow for modern ships, its competitiveness was blunted. Where it was already left behind by other docks and ports it faced economic ruin.

Until the 1960s barge traffic continued to provide most work for the river's tugs. Alexander, Watkins, Elliott, and Gamecock concentrated on ship towage, but other tug operators - such as Gaselee and Knight - depended on the lighter trade. In 1953 between 80 and 90 per cent of cargoes coming into the Thames were unloaded into around 7000 lighters for onward shipment as high as Brentford. Then this business, too, began to shrink as goods came in containers to Tilbury, where it was transferred to the road. Consequently the upriver warehouses and cold stores closed down. Coal for power stations went by road too, and oil was piped from the riverside terminals, or sent in self-propelled barges. Of the traditional cargoes, only London's refuse still left the city by barge, behind a tug. The result was the virtual disappearance of the Thames lighterman: there were 7000 in 1945, and 250 forty years later. Only on the Medway did freight lighterage survive.

Because London's trade never regained its prewar peak, the Thames very soon found itself with too many tugs. Rationalization set in much earlier than on the Mersey, but much less effectively. In 1950, Watkins with their current fleet of eleven tugs joined forces with Elliott's two boats to form Ship Towage (London) Ltd, under the chairmanship of Elliott's Colonel A. J. Page. Watkins had more work, but Elliott had the better tugs. Ship Towage then acquired the four Gamecock tugs from the Ocean Salvage & Towage Company. The company managed the tugs of all the constituent operators, forming a confederation in which each remained a separate entity, partly to preserve the goodwill of old customers, but mainly because none would compromise its independence. Together they were supposed to rationalize building programmes and operations so as to avoid wasteful duplication and competition in a reduced market, and by such economies, create resources for the modernization that would attract new customers to the river. In practice they could never agree on policy: each company modernized independently, if at all, and continued to compete with the others as if nothing had happened.

In 1960, in order to centralize its maintenance and repair facilities, Ship Towage acquired the ex-Admiralty workshops and dry docks at Sheerness on the Medway. They were run by a separate company, The Medway Dry Dock & Engineering Company. Five years later Ship Towage bought the four Gaselee ship tugs, and took over the running of that side of the Gaselee business. Gaselee retained their barge tugs. J. P. Knight still had seven barge tugs on the Thames in 1960 and a work-sharing arrangement with Sun Tugs, but six years later linked up with Gaselee to save overheads.

In 1968 Elliott's 'Vanquisher' with 1800 b.h.p. diesel engines was the most powerful tug on the Thames. Sank in January 1976 while handling a container ship 'Jervis Bay'. Salved and returned to service. (M.J. Gaston, Dunton Green).

W. H. J. Alexander's Sun tugs were now the group's biggest rivals. Ship Towage had more resources, but Alexander was better run. In 1969 they, too, merged with Ship Towage, bringing another fourteen tugs into the joint fleet. A new company was formed, London Tugs, in which control was vested equally in Alexander and Watkins. It held together a group that now had a near-monopoly of Thames towage: its writ ran everywhere except in the docks, where worked the little tugs of the Port of London Authority. But bigger was not noticeably better: there was still none of the efficiency and economy that only a single guiding hand could bring. Tugowners were a breed too independently minded and distrustful of others to work together for long. Furthermore, the postwar growth of tug business on the river had petered out with the closing of the upstream docks, the thinning-out of the passenger liners in the 1960s, and the growing size but smaller number and increasing manoeuvrability of cargo ships. A modern container ship or bulk carrier did the work of three older ships. Tankers of the 1950s had needed up to half a dozen tugs to move them, but by now the usual number was two or three. Business peaked in 1966; thereafter, revenue declined.

So much of the business of Thames tugs now lay with big tankers that their new tugs tended to be more powerful than those of many other ports. Elliott's *Vanquisher* of 1955, with her 1800 hp diesel engines, was the most powerful tug then on the Thames. Their *Avenger* of 1963 and 2250 hp was similar to the 2000 hp diesel Gamecock tug *Moorcock*.

William Watkins, however, retired deeper into conservatism and the rear ranks. Their few new tugs had inauspicious histories. The second *Cervia*, built for them in 1946, was sunk while undocking the liner *Arcadia* in 1954; five men were drowned. Watkins made a last attempt to regain a foothold in deep-sea salvage work with their latest *Rumania*, of 592 tons and 1300 hp. Bought in 1946, she was wrecked in the Thames Estuary ten years later. Watkins were slow to build diesel boats. Their first, the *Dhulia*, came in 1959. The penultimate Watkins tug perpetuated a famous name - *Hibernia*. Completed in 1962, she was also the company's most powerful boat ever, with a 2250 hp diesel engine. The very last Watkins-built tug was the latest *Burma*, put into service in 1967. With her ended a famous line that went back further than any surviving British tug business. A consequence of Watkins' inability or unwillingness to modernize may be discerned in the declining fortunes of two of their old tugs. In the year October 1953 to September 1954, the *Kenia* averaged seventy-two jobs per month. Between October 1963 and September 1964, the similar *Gondia* did only fifty per month, a drop of nearly a third.

By contrast, when London Tugs was founded the Sun tugs were the most formidable element in it. For enterprise and confidence in the postwar years, no operator bettered Alexander on the Thames. As early as 1946 they had replaced wartime losses with two 1030 hp oil-fired steam tugs. They began to go over to diesel very early, with the *Sun XVII* of 1951, which had bridge control. By 1953, Alexander had nineteen tugs on the river, including the biggest fleet of ship tugs. As always, their variety enabled them to handle all kinds of trade - from 1600 hp for the biggest visitors, down to 350 hp barge tugs. All had radio telephones, and the ship tugs had radar, firefighting, and salvage equipment for coastal and cross-Channel towage. Ten years later the first of a series of three boats, the finest and most powerful on the Thames and Alexander's swansong, joined the fleet. She was the *Sun XXV*, with 2400 hp and 22 tons of bollard pull.

In the meantime, while still an entirely independent operator, Gaselee had made a move heavy with significance for the future. The port of Felixstowe on the Suffolk shore of the Orwell had been a satellite of Harwich opposite, which in turn had been a satellite of London since 1886, when the newly-formed Felixstowe Dock & Railway Company opened a dock mainly serving coastal traffic carrying agricultural products. But trade forsook it after the First World War, and by the 1950s the port had become derelict. Then came renaissance, when new owners with an eye on its excellent position for cross-North Sea trade began to revive the port. Felixstowe, like other East Coast ports, was in a sense midway between Britain and the Continent, a position of special significance when Britain entered the Common Market in 1973, and trade restrictions with other member countries were lifted. In particular, Felixstowe benefited from its location opposite the port of Rotterdam: the Hook of Holland was only 110 miles away. Goods transhipped here flowed into Britain through the east coast ports in feeder ships.

Shippers also came to Felixstowe because it had no history, no burdensome past of outdated facilities, inefficiency, and bad labour relations. Management was modern and forward-looking, and the dockers bore no load of old prejudice against cost-effective cargo handling techniques. From the start, the labour force consisted of a small body of local men working full-time in pensionable jobs, and better paid than most dockers.

Trade passing through the port reached 376,000 tons in 1964, when an oil jetty was opened. There began a major programme of reclamation of land, on which modern harbour facilities were built. In July 1965 container berths were opened, to attract ships from Holland and Belgium. But Felixstowe's real boom began with the opening in 1967 of the first section of a purpose-built, deep-water container terminal, the earliest in England. With 33 ft of water, the Landguard Terminal could take 15,000-ton transatlantic container vessels direct from the United States. In its first year the terminal handled more than 18,500 containers. The total tonnage of freight handled by the port in 1968 was 1.5 million tons. By the end of 1974 it was 3.8 million tons and rising, a tenfold increase in the ten years since the transformation of the port.

Gaselee saw the potential early on. In November 1963 they stationed their ship tug *Hooligan* at Felixstowe, followed quickly in the same year by the *Agama*. Both were operated by Gaselee & Son (Felixstowe) Ltd, who opened an office with a manager in 1964. As trade increased, Gaselee and their successors put progressively more numerous and more powerful tugs into the port. To man as well as manage them locally, ten Thames tugmen and their families were encouraged to move to the area. After the formation of London Tugs, that concern acquired a half-interest in the Gaselee Felixstowe operation, and the composition of the little tug fleet came to reflect the new

'Burma', 1967, was the very last Watkins-built tug. With her ended a famous line that went back further than any surviving British tug business (M.J. Gaston, Dunton Green).

mix. By 1974 there were three tugs at Felixstowe - the Gaselee boat *Sauria,* Watkins's *Ionia,* and Alexander's *Sun XVIII.*

In Southampton and the South Wales ports, as in the rest of the country, national economic crisis in the mid-1970s ushered in a long spell of fluctuating fortunes masking a general decline.

From 1974, rocketing prices reduced oil consumption in Britain, hitting the tanker traffic in and out of Southampton. The container trade grew by 9 per cent every year for five years up to 1980; but from early in 1981 a running series of disputes, involving some or all of Southampton's 1300 dockers, restricted the container berths to a third of their normal capacity for eleven months. Well before they ended, in October 1981, Southampton's biggest container ship operator reckoned that it had already lost £20 million. By the time the last dispute was settled, in January 1982, container traffic was down by nearly half on 1980, and the port authority had lost more than £15 million in revenue. A few months later the Falklands War brought a temporary boost in business for Southampton and its tugs; but although the container traffic that had been diverted away from the port returned, it was slow to do so. Southampton sank from first to second place among

Britain's container ports. The port authorities proposed to put the docks on an economic footing by cutting their workforce by a third. The response was disruption that lasted from October 1984 to February 1985, further harming Southampton's reputation, and the revenues of its operators and customers. By November 1984, almost half of its container traffic had gone. Once more the customers came back, but slowly; and other ports benefited. And once more, providence in the shape of other peoples' troubles intervened to restore some of the port's fortunes. The year-long miners' strike of 1984-5 brought in more tankers, to keep Britain's oil-fired power stations working at full capacity.

For the South Wales ports 1974, too, was a bad year; particularly for those reliant on oil. Soaring prices reduced demand, causing oil traffic to drop. The freight handled at Swansea fell by more than 10 per cent, compared with 8 per cent for the other ports. Ship movements - and therefore work for tugs - declined throughout the 1970s. About 160 tankers called at Swansea in 1976; the number in 1977 was fewer than 100. Again, there were years when oil imports (though not exports) were up; but the new trend was remorselessly down. Llandarcy stopped refining in 1980 - an event which, on its own, halved the port's exports - and was used only for the storage of refined products. These were brought in by a smaller breed of tanker than the VLCC, that might not need tugs at all. Compounding Swansea's problems were falls in demand for steel exports, its third main prop, and for timber imports. Export customers were not encouraged by a prolonged and very damaging steel strike in 1980. This, combined with the halt in refining at Llandarcy, caused the port authority's profits to collapse from £2.1 million in 1979 to little over £500,000 in 1980.

Swansea set out to attract a greater variety of business; but this was uphill work, for it could not accept the largest container ships. The King's Dock led into the Queen's and Prince of Wales Docks, and its entrance was too cramped. There was accessible accommodation for the smaller sort of container ship and for traditional cargo vessels; but this was not the stuff of which major ports were made. Nor did it help towage operators, for smaller ships were able to manoeuvre with fewer tugs. Symbolic of Swansea's decline as a port was the metamorphosis of the South Dock into a recreation area. Wherever there was a ruined port, local enterprise was turning docks into marinas, or filling them in and grassing them over.

In the early 1970s Alexandra were in the throes of a wholesale and fundamental reconsideration of their role, necessitated by the same national crisis that had hit their outports. For the company, developments at Southampton and Swansea had been, broadly, the continuation of an old story. Southampton, in particular, was at times of critical importance to

Alexandra - for instance, in 1975 the container terminal there was the company's biggest single source of revenue - but in the 1980s, as in the 1920s, conventional tug operations in both ports would weather trade vicissitudes by exercising imagination and making sound business decisions.

Elsewhere, the Alexandra story was very different. The Company had lived through two great crises in its history - in the 1890s and 1930s - by boldly spending its way out of trouble (an option open to well-managed and properly financed tug operators, and therefore rare), but basically always in the context of ship towing in traditional ports. If that role was exploited with forethought, the crisis passed, because national economic revival had always brought prosperity back to the ports. Not so in the late 1960s; and not, in particular, from the early 1970s, when long-drawn-out national economic crisis added its burden to a British shipping industry in catastrophic decline, and when the trends of ship design were reducing the need for tugs. Alexandra's Southampton and South Wales offices soldiered on; but in Liverpool, the heart of the company's operations, the long term offered no real expectation of revival on the old scale. Clearly, with the traditional towing business in decline, other kinds of business must be found. Diversification would also provide a buffer against hard times: as the ports had discovered, these hurt less if your repertoire was wide. At the same time, conventional ship towage must not be neglected: every chance had to be seized. Faced with their third major crisis, Alexandra rose to the occasion in a way that the tough old Poulsoms, Bickets, and Macks of yore would have approved.

For the bigger, and it was hoped better, Alexandra that had emerged from their takeovers the years 1969-73 brought two completely new initiatives. One of these was a departure in more than one sense. In 1972 two businessmen from Gibraltar came to Alexandra to press the advantages of a presence in the colony. One Shell tug and three Royal Navy tugs handled all the traffic in the port and naval base; but their own employers had priority. There was more business from visiting tankers, cruise liners, and cargo ships than they could absorb; and they alone could not possibly deal with the salvage work outside. The venture was not quite so outlandish as it might seem at first sight, for Gibraltar was secure under British rule, for all the pretensions of Madrid, and its government and its people were enthusiastically pro-British. Because everyone spoke English and labour was cheap - local tugmen's wages were not much than half those paid in Britain - the prospect of hiring local crews under British masters was attractive. Finally, Gibraltar would give Alexandra a presence both in the Mediterranean and off the Atlantic shores of western Europe. Accordingly in January 1973 the *Wapping* (the former *William Lamey*) moved in from Swansea, and immediately produced results. By April, she was averaging between six and seven jobs a week.

When Alexandra moved into Felixstowe they had 'Sauria', 'Langton' and 'Egerton'.

'Canberra', home from the Falklands, receiving a rapturous welcome ashore and afloat 11th July 1982.

The other new initiative of these years had a more bumpy start. In anticipation of the opening of the Seaforth grain terminal, due in 1972, Alexandra began to build up their own barge fleet for the first time. (Experiments with concrete barges in the 1920s had come to nothing.) In 1971 they acquired Bulk Cargo Handling Services, a business which in turn bought the thirty-one barges of the Liverpool Lighterage Company, of which Paul Mack was already a director. Towards the end of 1972 the Rea Towing Company dropped their lighterage business, and Alexandra acquired their barge fleet, together with one of their barge tugs. In 1973 BCHS also bought the seven elevators of United Grain Elevators, rationalizing Alexandra's grain trade activities. The company placed an order for three 1,000 ton, self propelled grain barges, for a total of £420,000: the *Seacombe Trader, Sealand Trader,* and *Seaborne Trader*. Unfortunately, long delays in the opening of the Seaforth grain terminal meant that no revenue was forthcoming from these costly investments until November 1974.

Meanwhile the company also pursued and exploited every traditional towing opportunity. In particular they expanded into towage in coastal and home waters as never before. At first Alfred Lamey, who had remained on the board of his company when it became a wholly-owned subsidiary of Alexandra, had responsibility for developing this side of the business. The famous Lamey competitiveness and initiative that had earned the family their Jolly Roger now found a constructive outlet on behalf of their old rivals. As subcontractors for Rea, Alexandra tugs towed barges from Bantry Bay to the Esso refinery at Cork. Long coastal tows of big, modern ships, for which Alexandra tugs had not been particularly well suited, were now sought. When the powerful new tug *Mumbles* was not needed at Port Talbot, she did jobs such as the tow of a ship from Milford Haven to Antwerp in Belgium. Two Alexandra tugs towed a liner from Barrow-in-Furness to Newcastle.

The work Alexandra lost was almost as significant as that which they gained, for it demonstrated the strength of the competition they faced, and showed how enterprising and persistent they had become. Between 1969 and 1974 Anglesey again became of great interest to Alexandra. A new ore terminal was built at Penrhos by Holyhead, with work for barges, dredgers, and tugs. By 1970 there were contracts for towing aluminium oxide carriers of up to 55,000 tons, which locals, not Alexandra, won. But when in 1973 and 1974 Shell was building a new harbour at Amwlch, on the north coast of Anglesey, Alexandra's *Alfred* and *Crosby* were at different times standing by the dredger *Beaver Chief* as she dug a trench under the Menai Straits for the pipeline connecting the Rhosgoch terminal, inland from Amwlch, with the Shell refinery at Stanlow on the Mersey.

As in earlier crises, Alexandra knew that money must be spent on keeping their fleet up to date if the available business was to be kept and new work won. The new emphasis on sea towing influenced the design of additions to the fleet. The *Margam*, sister to the *Mumbles*, was typical - a big, powerful, sea-going tug based with the *Mumbles* at Port Talbot to handle ore carriers, but intended from the start for sea work as well. In 1970, the year she joined, four more tugs were ordered from Richard Dunston, all with around 40 tons of bollard pull and 2200 hp, all equipped with Kort rudders, and each costing £230,000. The first pair, the latest *Alfred* and *Crosby*, was delivered in 1971, and the *Albert* and *Victoria* during 1972. Fortunately, Alexandra and other tug owners could continue to rely on Government and bank finance for this essential growth: there was not enough revenue to allow the building programme to be self-financing, and probably never would be again. Meanwhile, between 1969 and 1973, around twenty older boats were taken out of service. One was the *Canning*, sold in 1973. The company's last steam tug, she went to the Swansea Maritime Museum in 1974.

The figure for tugs sold or scrapped reflected more than just a policy of replacing old tugs with new. Behind Alexandra's success in finding new work in new and old areas lay a battle against ever-growing difficulties of all sorts. There was less and less conventional work for their tugs to do, particularly in Liverpool. There the Alexandra fleet, which had numbered thirty-three boats in 1968, was down to nineteen in August 1971. Rea had nine tugs, and since they had most of the tanker traffic, they were doing better. Most of Alexandra's tugs were the surviving Cock and Lamey boats. Since 1970 they had sailed in Alexandra colours and under Alexandra names; the 'cock-a-doodle-do' Cock tug whistle signal (one long blast, four short) was heard no more on the Mersey.

For the present, no tug was on call round the clock. There were Esso and BP tanker towing contracts for Alexandra; and the Royal Seaforth Container terminal opened. But tankers and container ships notwithstanding, fewer ships were coming to the Mersey. In 1961 the number had been 34,000, ten years later it was 28,500; a steady drain, averaging ten ships a week. In 1972, the year in which Seaforth was opened, the whole southern docks system was closed down. In spite of rate rises generally once and sometimes twice a year, Liverpool income fell consistently. Typically, the first eight weeks of 1971 showed a decline of 480 jobs over the same period in 1970. In 1972 the figure was down by 265. For 1973, work in Liverpool was down by 2400 jobs compared with 1971.

As tugs went for sale or scrap, so crews had to go as well. A redundancy programme implemented in August 1970 was intended to reduce the strength by thirty men, or six crews. In fact, more than fifty men took the terms offered. To them, the towage business seemed to offer no future. Four more crews were made redundant a year later, another four in November the same year, and a further four in 1973.

As always, one potent cause of falling revenue, tug strength, and staff

'Canada' assists the biggest ship ever to enter Liverpool Docks. 'Atlantic Compass', 57, 255 G.R.T. 292.52 metres — just one metre shorter than the QE2. She arrived from Nova Scotia in a north westerly force seven wind and slipped smoothly through Gladstone River Entrance onto her berth at Royal Seaforth Container Terminal, 1987.

The first three tugs built for the Thames in 1977 after the acquisition of London Tugs in 1975. 'Sun London' with 45 tons bollard pull, 'Sun Essex' and 'Sun Kent' with 35 tons.

numbers was labour trouble, during which work stopped coming. A pay dispute brought the Liverpool tugmen out for six weeks in April and May 1970. There was a national dock strike in July and August 1970, and another two years later.

There might now be less competition on the Mersey, but the spread of the Cory empire made it more threatening now that Alexandra was working so much outside their traditional ports. In 1970, R. & J. H. Rea took over Steel & Bennie of Glasgow and the Newport Screw Towing Company. Two years later the Grangemouth & Forth Towing Company was acquired by Cory and the Clyde Shipping Company; and the Plymouth tug operators W. H. Reynolds went to Cory in 1973.

Business, and revenue, at the other Alexandra ports was generally better. Late in 1971 it was noted that Southampton and Swansea were the company's only financially viable ports. Wages there had to go up, and as at Liverpool, rates had to rise to compensate. In 1971 a Port Talbot ore carrier cost £300-£400 to berth, according to size. But in contrast with the Mersey, this automatic interaction of wages and towage rates did not happen every year. At Swansea, six tugs went in the three years 1969-71. Falls in revenue accounted for some losses, but others made way for the fewer, larger tugs needed to cope with the Port Talbot ore carriers. Southampton had a dock strike of its own in 1969, and both ports suffered in the two national dock strikes; but otherwise they were trouble-free. The prospects at Southampton were sufficiently encouraging to persuade Alexandra to start buying the shares of Red Funnel, a public company. Between 1972 and 1976, a little more than 10 per cent of the equity was acquired.

With such mixed fortunes, Alexandra's overall revenue dropped a little almost every year - by 4 per cent in 1969, 2 per cent in 1970. It rose in 1971, then fell back in 1972 and 1973. In 1972, the only side of the business to show an increase was coastal towing. At the same time, outgoings inevitably climbed. With interest payments, a policy of constant renewal initially cost more than the savings effected in the repair bill. Fuel costs fluctuated, but even at its low point in November 1971 fuel oil cost at least £10 15s a ton, or 170 per cent more than in 1963. Inflation had already begun.

Partly as a result of the increase in sea towing work, and partly because ship towing business at their traditional ports was so inconsistent, Alexandra's tugs were no longer tied so closely to particular home ports as hitherto. They and their crews could expect to be moved around from job to job - some in locations strange to Alexandra - as the work dictated. Sometimes, it seemed, they might be called upon to go very far afield indeed: in 1973 Alexandra was taking an interest in the development of Saldanha Bay in South Africa, where a new harbour on Port Talbot lines was to be built. Although no Alexandra tug went there in the end, their interest in such a distant project was symptomatic.

In 1967 a portent had floated into the Mersey: a sign that pointed a way out of the late 1960s doldrums. It took the ungainly form of the oil exploration platform *Bedford X*, towed in from Barrow by Alexandra tugs.

Work for tugs in the British offshore oil and natural gas industry began with the birth of the industry itself in 1964, when licences to drill for both sources of energy were granted by the Government. Success came first to North Sea gas: the first wells were drilled in the same year, and the earliest major strike was made in 1966. Oil drilling began in the British sector of the North Sea in December 1964. There, the earliest discovery of oil was made by Amoco in 1969 - the small Montrose field. Other exploration began in the Celtic Sea - the Atlantic waters bounded by Southern Ireland, Wales, the west of England, and Brittany.

The first major oil strike came in Norwegian waters, after forty holes had been drilled in the sea bottom: the Ekofisk field was found in November 1969. For Britain, the first big discovery was the Forties field, located by BP in October 1970. The craft that located it was a new sort of 'ship' - the submersible drilling barge or exploration platform *Sea Quest*. In 1975 the Forties field came on stream. In 1974 the Ninian field far to the north-east of Shetland was discovered. Production began in 1978, the oil flowing ashore to the Sullum Voe terminal 109 miles away. To Sullum Voe, too, came oil from a new strike, made by BP in 1980 - the Magnus field. Another terminal sprang up at Flotta, on Scapa Flow in Orkney. At all such terminals, powerful tugs were needed to move the tankers that came to collect the oil. In 1975 the offshore oil reaching Britain totalled 1.5 million tons; in 1982 the figure was 103 million tons. The North Sea boom was in full spate.

It was natural that east coast tug owners, particularly those with deep-sea boats, should be early in the field when North Sea oil exploration got under way. United Towing's 5000 hp *Englishman* of 1965 was the biggest tug ever built for British owners; but only six years later the company's fleet included the *Lloydsman* of nearly treble the power, the 12,000 hp *Statesman*, and three tugs of the *Englishman*'s capability. These could handle oil rigs as well as the biggest bulk carriers that came to the Humber. Oil exploration platforms were built or fitted-out on the Tees from 1965, making work for Crosthwaite.

After the mobile exploration platforms came the much more massive permanent production platforms. Like the exploration rigs, they were built in existing ports such as Middlesbrough, Barrow, or Birkenhead, or in the new fabrication yards that sprang up in the deep-water sea lochs of Scotland, like Nigg Bay on Cromarty Firth, Ardesier Bay on Moray Firth, Methil on the Firth of Forth, or Loch Kishorn. A production platform's core was its jacket - a steel or steel and concrete structure that had first to be towed from one construction site to another for fitting-out, like a ship. Then it was towed out to the oil well, where its legs were flooded and it was upended and

Arriving or departing, 'Sun Essex' was on hand to help.

Float-out of 'Elf 1' in 1976 destined for the Norwegian Frigg gasfield, an operation that calls for a high degree of skill and co-operation between tug skippers.

sunk over the wellhead. A typical jacket weighed around 40,000 tons, but they could be twenty times the size, the largest moveable objects made by man. When complete, they towered perhaps 500 ft out of the water. A team of modern, powerful tugs could move a jacket around the construction sites in sheltered waters, and out to the seaward boundary of a port, as Crosthwaite tugs did with the first jacket completed on the Tees in 1974; but there the specialized ocean-going boats of United Towing or foreign operators had to take over.

Once the jacket had been sunk in position, the elements of the superstructure of each production platform had to be floated out to the site, lifted to the top, and assembled. These included building units for drilling and control machinery, workshops, generators, accommodation for the rig crew, the helicopter landing pad, and the derrick itself. Finally, the pipeline to carry the oil from the production platform to its land terminal had to be laid. Tugs were central to these operations too. They were needed to tow out and manoeuvre the big construction barges that carried the rig superstructures, and the heavy-lift crane barges that raised their component parts into position. They were needed to tow the pipe-laying barges, not all of which were self-propelled. And after either an exploration platform or a production platform was in place, an entirely new sort of boat would be needed to service both: the support vessel, or anchor-handling tug.

These big sea-going boats, recognizable from their disproportionate area of open cargo deck aft, combined the roles of supply ship, ferrying supplies and equipment in and out of the fields, with that of moving the exploration rigs. These floated on submerged pods while they drilled, tethered to the sea bottom by eight or nine anchors, each on the end of up to a mile of chain. When the platform was moved from one exploration site to another, the anchor-handling tug was called in to winch up the anchors, each weighing between 15 and 30 tons, together with their marker buoys. Each anchor took an hour to raise, in calm conditions. The tug then towed the platform to its new site, and laid out its anchors again.

In 1973, Alexandra began to focus their attention on the oil fields. The *Huskisson* went to work at the Flotta terminal in Orkney, where she remained for two years towing pipe-laying barges. Revenue was down 7 per cent in the year, and inflation was a menace, but the prospects looked good. At the same time, the difficulties of all tug owners were about to get worse, for the Arab - Israeli war in October 1973 plunged the Western world into an economic crisis unequalled since the 1930s. Since 1969, Alexandra had battled hard to hold their ground, in already desperately adverse circumstances, meeting with no worse than mixed fortunes. The dangers were suddenly greater, but so were the opportunities.

Among those who would bear the brunt of the test, a few faces had changed. Harold Mack retired in 1972, Henry Bicket replaced him as chairman, and Graham Mack was appointed managing director (finance). One new face was that of Captain R.J. (Dick) Braithwaite, formerly of United Towing, who was engaged late in 1973 as General Manager (Marine) - the man who on the retirement of Alfred Lamey in June 1974 took charge of day-to-day operational control of the company's excursions into wider waters.

During 1974, there began the public metamorphosis of Alexandra from a conventional towing company still, for the most part, relying on conventional trade, into today's many-sided group of transport businesses.

By autumn 1974 the Sullum Voe terminal had begun to take shape, and and in anticipation of the 300,000-ton tankers that would come, the Shetland local authority asked for four Voith-Schneider harbour tugs fully equipped for oil terminal work. Alexandra's approach was rejected, because others had got in earlier and Scottish tugs were favoured; but construction barges and tugs to tow them were also needed, and here Alexandra won their first such contract. Then, in 1976, the *Crosby, Alfred,* and *Albert* towed a pontoon barge from Nigg Bay to Moray Firth for oil platform construction. The same year, Alexandra helped at two 'float-outs' of production platform jackets. The first was in March, when the platform *Elf I,* destined for the Norwegian Frigg gas field, had to be towed thirty miles from its fabrication yard on Rothesay Bay to its fitting-out anchorage in Loch Fyne. The other float-out was at Loch Kishorn, where the steel and concrete jacket for the Ninian Central field was to be moved four miles into Loch Carron for further work. It was destined for what was then the world's biggest production platform: the jacket alone was more than 150 yards across, and weighed 154,000 tons. The tugs called for were the *Alfred, Crosby,* and *Albert,* plus the *Margam* from Swansea.

There was plenty of other coastal work - or 'offshore', as it was increasingly called. Barges and dredgers were towed from Rosslare to Stornoway; and six pontoon sections from Dublin to the Mersey to make a new Liverpool landing stage. In August 1976 the *Mumbles* towed Britain's largest floating crane - the *Brunel* - from Dunkirk to Narvik in northern Norway, then to Nigg Bay, where she was needed to lift a sunken 1200-ton Admiralty tug. Being one of the two most powerful boats in the fleet, the *Mumbles* was much used for long sea tows, including one from Falmouth to Bilbao, and another from Portsmouth to Valencia. A company distance record was set two years later by a new tug, the *Waterloo,* which took a dredger from Rotterdam to Haifa in Israel.

The *Waterloo* was one of a big crop of new tugs that had become essential to deal with Alexandra's expanding offshore business. No fewer than six new boats went into service in 1977, all built by Dunston, and all with navigational, salvage, and firefighting equipment that fitted them for most major tasks at home and abroad. The *Waterloo* and *Wellington* had a bollard

This is an aerial shot taken during the float-out of 'Elf 1'. The wing of the Alexandra Aviation Aztec is just visible at bottom left (Liverpool Daily Post and Echo Ltd.)

'North Isle' and 'Romsey' with the controversial nuclear powered vessel, U.S.S. 'Savannah'. (Beken & Sons, Cowes).

pull of 54 tons, and the *Wallasey* 45 tons. The other newcomers - *Sun London*, with 45 tons bollard pull, and *Sun Essex* and *Sun Kent* with 35 tons - bore names strange to Alexandra, but with a familiar enough ring on the Thames. They were the first fruits of another change of direction for Alexandra: this time south-eastwards.

By the mid-1970s London Tugs were ripe for takeover by a progressive management that would finally bring unified control and true, long-overdue rationalization to the Thames. At this time only the Medway Dry Dock & Engineering Company was making money, a considerable achievement for a ship-repairing business when Britain's merchant fleet was in headlong decline. William Watkins, a director of London Tugs and the last tug operator to bear that famous name, asked if Alexandra would be interested in acquiring the company. So it was that in late 1974, a 'foreigner' from Liverpool took London Tugs over lock, stock, and barrel. The red and white banded funnels of the Sun tugs, together with all the river's other distinctive liveries, vanished into uniformity. A new company, The Alexandra Towing Company (London) Ltd., moved into a new building on the site of Sun Tugs' old headquarters at Royal Terrace Pier, Gravesend. The transformation of the Thames towing scene was complete. Alexandra themselves were scarcely less changed, with their fleet doubled in size overnight to fifty-four boats. This was 'expand and survive' with a vengeance. Furthermore, when Alexandra took over London Tugs, they took over not only all the ship towing on the Thames, but also all that of the port's most important satellite. They acquired the half-share of Gaselee & Son (Felixstowe) still in Gaselee hands, and substituted the Alexandra Towing Company (Felixstowe). The three tugs working there in 1975 were the *Sauria*, the *Deben* (the *Sun XXII*, renamed after a Suffolk river), and Alexandra's own *Egerton*, all under the direction of Dick Hazell, formerly of Gaselee.

Fifteen more tugmen's families moved from the Thames to the Felixstowe and Harwich district. This did something to cool a sense of grievance that the Gravesend tugmen bore towards those at the Suffolk port. Crews working at Felixstowe and Harwich made more money than Thames crews, whose trade was in slow decline. Since both ports were regarded as satellites of London, where their work traditionally belonged, resentment was added to the envy. The rift twice brought towing on the Thames to a halt. In an attempt to satisfy the remaining Thames crews, Alexandra's Gravesend tugs took turns working at Felixstowe, two at a time, so that both fleets could share in the glowing future of the company's latest outport.

In 1976 Alexandra bought a share of ITM Offshore Ltd, a Teesside offshore operator busy in the North Sea. Three years later the Alexandra Marine Transportation Company was founded to take over the operation of all Alexandra's offshore interests with R. J. Braithwaite as Managing Director.

The *Wellington* and *Waterloo* quickly found charters of at least six months in the North Sea, and in summer 1978 the *Wellington* went to Sullum Voe for three months. Long jobs such as this called for fast crew changes away from home; a need that first became apparent during the two-year contract at Scapa Flow. To satisfy it, in 1975 Alexandra had made another radical new departure. They bought a five-seater Piper Aztec aircraft, christening it *Flying Kestrel*, and formed Alexandra Aviation. Another aircraft, a nine-passenger Britten-Norman Islander, was bought in 1977. In October 1979, Alexandra acquired Euroair Transport to operate their growing new fleet, which by 1980 numbered six planes. As well as passengers, the bigger aircraft could carry spare parts and stores for tugs.

Alexandra's expansion and diversification was in the nature of a never-ending race against economic pressures and keen competition, a race in which, gradually, they drew ahead. From 1975, Cory - now part of the Ocean Group - were also building modern tugs in quantity. No fewer than ten were ordered that year, bringing the Cory fleet strength up to over fifty. In 1978 three new Cory tugs, operated by Shetland Towage Ltd, became based at Sullum Voe.

The worst financial pressures were the direct or indirect result of the Middle East oil crisis that struck late in 1973, and brought on a recession that began to lift only in the mid-1980s. Its effect on tug fuel prices was terrifying. Fuel oil had been £18.79 a ton in 1973. In the first shock of shortage, in 1974, the prices went up to 160 per cent to £47.90, and by 1976 it had reached £67.97, or more than three and a half times the figure of three years earlier. The age of truly cheap oil was over. Other costs mounted as part of a general inflation fuelled by oil prices. The bigger, sea-going additions to the fleet made in 1977 cost over £1.1 million each: a quadrupling of price in eight years.

Alexandra's old ports were in no condition to help win the race, with the exception of Southampton. Primed by towage rate rises in most years, including two in 1978, revenue was encouraging throughout the 1970s, and the tugs went on making profits into 1981. From 1975, Liverpool's trade pleased everyone by steadying for three consecutive years. Early in 1976, Alexandra had lost only three tugs from their Liverpool fleet in 4½ years, and its strength now stood at sixteen. In the climate of the time, a slow-down in decline was hopeful progress. One factor that helped was the withdrawal from Barrow-in-Furness of the Rea tugs that had been stationed there, leaving Alexandra to provide all the assistance. There were rate rises almost every year, prompted here by equally regular wage rises. Rea lost no tugs at all, and had enough business to allow twenty-four hour working.

Revenue at Swansea and Port Talbot fluctuated as readily as their trade, that from Port Talbot suffering particularly from the cutbacks in South Wales steel production in the mid-1970s. In spite of massive rate increases, the fall in the number of tankers calling at Swansea in 1977 caused two tugs to be

Another hazard for tug men is fire-fighting. 'Bramley Moore' is seen here during an exercise on the Mersey giving the 'Northia' a drenching (Liverpool Daily Post & Echo Ltd.).

'Pacific Glory' on fire after being in collision. Alexandra and Red Funnel tugs combined to send vessels to her aid (Chris Wood, Southampton).

taken out of service in December, and eighteen tugmen to be made redundant. The fortunes of the Gibraltar operation were as fickle as those of the South Wales ports. In September 1975 the *Wapping* was relieved by the *North Isle*. Trade was not yet consistent enough to make work for more than one tug. On the Thames, Alexandra put up towage rates for their newly acquired fleet twice during 1975, to check the slow haemorrhage caused by falling business and rising wages. For good measure, they raised the rates at Felixstowe as well.

Difficulties notwithstanding, Alexandra's new initiatives gave them two brilliant and two satisfactory years between 1975 and 1978. In the first four months of 1975 overall revenue was up - the first rise in three years. This was an early sign of a record year, due principally to new revenue from the acquisition of London Tugs, with offshore work contributing another 20 per cent of income. More important, net profit for the year was a handsome £900,000; and for the first half of 1976 alone, it was £1.1 million. After the death of Frank Batey, killed by an IRA bomb, the Tyneside tug operator Lawson-Batey was offered to Alexandra, but nothing developed. The 1977 result was affected by Swansea's current downturn - but even so, revenue ended up higher in both 1977 and 1978, in each case producing 'a good result for a bad year'.

Salvage as a source of revenue had been patchy for everybody since the Second World War. Ships were becoming more reliable and, with the spread of navigational aids such as radar, safer too. In 1938 forty-eight British ships had been lost at sea; in 1960, the number had fallen to seventeen. As early as the 1940s, Tees Towing, with a monopoly of towing on their river, were getting only eight salvage jobs a year. At Gravesend by the late 1960s, the whole of the traffic on the Thames, from bulk carriers to barges, generated in a week perhaps one salvage job; a 'hovel', as it was called on the river.

The jobs that were there, made up in drama and variety what they lacked in numbers; and the revenue that they produced was individually higher - when they brought any at all. Captain Dan Parker of the Falmouth-based, 4000 hp ocean-going tug *Turmoil* and his employers, the Overseas Towage & Salvage Company, were among the unlucky, though their story provided the maritime sensation of 1952. On 4 January that year Parker picked up the crippled American freighter *Flying Enterprise*. For five and a half days he fought to bring her into port, finally failing on 9 January sixty miles off Falmouth when the freighter's split hull gave up and she sank. 'No cure, no pay' was the salvage agreement, so Parker got nothing for his efforts except a certain measure of fame. Even this went, for the most part, to the master of the *Flying Enterprise,* who insisted on staying with his ship when all others had left her.

Although there were more British ocean-going salvage tugs now than before the war, they remained few. Most of the time they waited about for work in their West Country or Irish ports, and so were grossly uneconomic to run. Most offshore salvage work was perforce done by ordinary harbour tugs, or by the bigger and more suitable coastal tugs when they became numerous. Alexandra's tugs participated in most of the dramatic rescue and salvage operations around Britain's coasts. The most spectacular and lucrative were tanker rescues, for the size of the ships made them difficult to handle, and their cargoes were both dangerous and valuable.

Late in the evening of 23 October, 1970 the 119,000-ton tanker *Pacific Glory* collided in darkness with another tanker, the *Allegro*, off the Isle of Wight. Holed, the *Pacific Glory* ran aground and burst into flames. In terror of the fire and explosions, her Chinese crew leapt into the oil-choked sea: thirteen of them died. Four Alexandra tugs out of Southampton, Red Funnel's *Calshot*, and later London Tugs' *Avenger* converged on the wreck. The *North Isle* and *Brockenhurst* got lines aboard the tanker fore and aft to hold her secure. By this time 'flames were belching out all over', and dead men were hanging over the stern rails. Braving the danger of more explosions, the *North Isle* went alongside to put men of the Hampshire fire brigade on board. At one point the oil in the water caught fire, surrounding both tugs in a ring of flames.

In weather worsening to a gale, the two tugs clung to their station from midnight on the 23rd to the evening of the 26th. Their crews described the night of Sunday, the 25th as 'hell'. Then, with the arrival of Smit tugs big enough to tow the ruined hulk into Lyme Bay, the *North Isle* and *Brockenhurst* limped back to Southampton, storm-damaged and with their crews exhausted. Meanwhile the *Huskisson* and *Coburg* had arrived from Liverpool, armed with the oil-dispersing gear that some tugs had carried since the *Torrey Canyon* disaster, and they stood by the Dutch boats during the tow.

The *North Loch* and the *Ventnor*, boats that had been involved in the *Pacific Glory* operation, salved the BP tanker *British Diplomat* in even filthier weather. It was so bad at their first attempt to reach the tanker that they had to turn back. At one point those aboard the *North Loch* actually lost sight of the *Ventnor* as she dropped into the trough of a wave. The tanker had lost her steering, so the *Ventnor* towed ahead, with the *North Loch* astern as 'rudder'. With the plunging and yawing of the casualty, the *Ventnor*'s tow rope parted, forcing the crew flat on their faces as it lashed back at them. The eighteen-mile tow to Cowes roads took between seven and eight hours. This job won the company and the two crews £11,500 to themselves. Struck by hurricane-force winds at her Tranmere moorings in January 1976, the 190,000-ton Shell tanker *Myrina* broke loose, drifted across the Mersey and grounded. Seven Alexandra and six Rea tugs refloated her, Alexandra alone collecting salvage of £57,000.

The bow of 'Eleni V'. Work had to stop when divers found they were welding directly over a tank full of oil (The Press Association, London).

Wheelhouse of 1960s vintage. Compare this traditional steering wheel and surrounding equipment with the photograph on page 267.

On 16 March 1978 the 234,000-ton *Amoco Cadiz*, wrecked off the Brittany coast, broke up and spilled 56,000 tons of oil. The *Sun XXV* and *Vanquisher*, by now part of Alexandra's Thames fleet, stood by her into early April to help prevent the pollution spreading to the Channel Islands. In May of that year, six Alexandra tugs with detergent-spraying apparatus and one United Towing boat went to the assistance of the *Eleni V* when she became a victim of the North Sea.

Oil exploration platforms were being built on the Tees, and if one broke loose from its moorings it could spread general mayhem in the river. In November 1965 the platform *Ocean Prince*, with thirty men on board, parted her cables at Smith's Dock in an 85 mph gale and ran down-river for two miles, bringing down an overhead power-line 200 ft above the water and colliding with a tanker before grounding. All the Tees Towing tugs were involved in trying to tow her back upstream, against the wind, but found themselves being blown sternwards. The tugs got some lines ashore, but they broke when the weather worsened. So did three towropes. This time the platform was blown across the river, ramming a dredger; then it was blown back, hitting a vessel laden with explosives. Finally the rig went aground again. When the gale eased, the tugs were finally able to tow her back to Smith's Dock, after thirty-four gruelling and dangerous hours. In 1967 another exploration platform, the *Staflo*, broke loose from her berth and drifted three-quarters of a mile down river.

Other salvages were equally untraditional, or even unique. One had its comic side. The barge *Gib How* was dumping stone at the site of the future Seaforth container terminal in the Mersey by releasing her 1000-ton burden through her bottom. On one occasion, perhaps through some miscalculation of the tide, she became possibly the first vessel in history to run aground on her own cargo. Nothing that the tug *Canning* could do would shift her.

In 1982, Alexandra tugs helped in the lifting of Henry VIII's warship *Mary Rose* from the bottom of the Solent, where she had lain for over 400 years. The heavy-lift crane barge *Tog Mor*, that was to raise the wreck in a specially fabricated cradle, was towed from Middlesbrough by the *Wellington*. At the site, barges into which cradle and ship were laid when they emerged from the water were manoeuvred into position by the *Victoria* and *Romsey*, which afterwards escorted them into Portsmouth. As a public gesture, Alexandra made no charge for their services.

In other areas, too, the twentieth-century tugman's life has also been quite as varied as that of his forebears. During the making of the film *Moby Dick* in 1954, Crosthwaite's *King's Cross* won a four-month contract towing a facsimile of the *Pequod* and two 75-ft artificial whales. One whale broke loose between Wales and Ireland and was washed ashore; the other was lost without trace. These waters were likely to lead tugmen into strange adventures. In 1922, during the Irish Civil War, Elliott's tug *Warrior* was hijacked by the IRA, who locked the crew below, substituted their own men, and intercepted a vessel carrying arms to the Government side. Tug and crew were later released unharmed.

In 1947 a young Norwegian radio operator was married on board the *Gladstone* by Swansea's Norwegian pastor, after discovering that formalities could be curtailed that way. Later a Swedish couple chartered the *Langland* for the same purpose. The logical outcome of this sort of activity duly arrived in June 1946. No wartime 'GI bride' sailing to America with her husband was allowed to travel if more than seven months pregnant; but one started to give birth prematurely, on board the troopship in Southampton Water. With a doctor, she was taken ashore on board the *Hornby*, but the baby arrived in the wheelhouse on the way.

Time was reluctant to deliver tugmen from the dangers of their trade. In the hard times of the 1920s, tugs had to be kept in service long after they should have been retired or at least overhauled. One of the achievements of George Robinson, in his thirty-one years as Crosthwaite's superintendent engineer, was to keep his employers' rickety fleet afloat. The average age of the tugs was thirty-nine years. They still included two single-cylinder wooden paddlers, and it was possible to see daylight through the planks of the *Granite City*. The *Florence*, of 1879, had no engine room telegraph to the end of her life in 1927: the skipper stamped once on the deck for the engine to start, twice for it to stop, and three times for the engineer to put her astern. If the planking was in the same state as that of the *Granite City*, this sounds a hazardous procedure.

The availability of navigational aids did not mean that they were used, even as late as the 1950s. Sometimes even medieval aids were lacking. Arthur Langton remembers when the *J. H. Lamey*, looking for her tow, was immobilized in fog and darkness somewhere off the coast of North Wales, totally lost because William ('Billy') Lamey had mislaid the boat's compass. He ordered the crew to line up along the bulwarks, listen, and sniff, on the principle that if they could hear sheep and smell grass, they were close to land. The tug avoided running ashore and a near miss by a passing ship, and when daylight came finally found her tow, an aircraft carrier.

Tug service was as stressful as ever, particularly for the officers who bore so much responsibility in their small commands, and Alexandra in 1960 instituted compulsory retirement at sixty-five for their 'floating staff'.

The ordinary lethal accidents of the trade have never stopped, in spite of modern tug design. On Christmas Eve 1984, Alexandra's *Implacable* was on her way to the Falklands on Ministry of Defence charter when she capsized in heavy weather in the English Channel, going down in eight minutes and drowning her chief engineer. Sometimes tugmen were lucky. While she was towing at Southampton, one of Alexandra's tugs collided with a Norwegian container ship. She capsized and, unable to release her

Moving forward in time from page 264, technology has brought the steering wheel down in size and introduced new equipment in a console on board 'Formidable'. Not to mention a seat for the skipper!

tow, was dragged. Remembering to shut the doors behind them, her crew jumped before she turned over. Their forethought saved the tug from sinking, and left one seamen standing on her keel, waiting to be rescued. The Russell family of Gravesend, Watkins tugmen, were accident-prone even by the standards of the trade. Harold Russell was one of the few to escape unscathed: his father was drowned in the *Badia* in 1926; a brother, William, was among those missing from the *Cervia* when she was girted by the *Arcadia* in 1954; another brother, Jack, was mate of the *Gondia* when he lost a leg, catching it in a mooring rope in the Royal docks; a third brother, George, was deafened for life by a blow from a parting hawser while towing a sailing ship, and during the Second World War, his ship was torpedoed. Harold Russell himself saved the life of a man the *Dhulia* nearly killed. Late one very black and foggy night in 1970 she was towing a pontoon five miles off the mouth of the Tyne when her towrope swept her mate, Stan Wood, into the sea. Dragged under the tow by the wash, he was bumped back along its keel. When he surfaced at the other end, he was completely disoriented, and began swimming in the direction of Denmark. Russell, the *Dhulia*'s master, had spotted Wood's legs disappearing over the side and switched on his searchlight. By chance the beam picked up the reflection of Wood's eyes. After ten minutes in the icy water he was back on board, little the worse.

For some years the tugman's rewards did little to compensate him for the risks and drawbacks of his trade. As a master with Ring Tugs on the Thames in the early 1930s, William Simmons might make as little as £323 in a year (on average a little over £6 a week) including all overtime and commission. With better times in the late 1930s, and a stronger union, tugmen's basic wages improved. A Watkins master was paid £5 17s, a mate about £4, deckhands £3 12s 6d, and the boy £1 2s 6d; while J. P. Knight's crews did better still. As before, wages went up fast when war broke out. A Liverpool dock tugmaster in 1943 got £7 14s, plus extra for Sunday working, war risks, and 'deep sea' work outside the port's limits. The officers did not get overtime as well, but the ratings did, as well as the other supplements. Alexandra's Southampton masters and engineers, who did not receive commission, were paid about 25 per cent more by way of compensation.

This time there were no wage cuts at Alexandra in the aftermath of war, for there was no general trade slump to occasion them. By 1960 an ordinary seaman was getting the Liverpool skipper's wartime basic pay. Ten years later, overtime usually doubled the basic wage for all ranks, without the need to work ninety hours a week to achieve it.

As far as his basic wage was concerned, the Liverpool deckhand of the 1980s was worse off than a docker, so in this respect their positions were reversed when compared with those of the 1890s, but if overtime were counted, less had changed - the tugman was doing better than the docker, as he had a century before. This was the factor that, for more than half that period, had been essential to any realistic comparison. Salvage payments were still a significant element in a tugman's income - to the extent that before a tugowner accepted one, the amount and the crew's share were put to them for approval; and crews on their own account could, and did, appeal against court awards.

The progression by which a school leaver became a tugman and rose through the ranks was unchanged. There were simply more ranks; or rather, a subdivision of ranks, dependent on ability and experience and defined by pay limits. A boy was the 'boy' in Alexandra until 1969, after which he started as the most junior deckhand or ordinary, unskilled seaman. Acquiring service and experience, the deckhand was rated as able seaman; then rose to mate and captain if he was good enough. If he was of a technical turn of mind, he went below decks as an engineer and then chief engineer. With the coming of oil firing and diesel tugs, there were, of course, no trimmers or firemen. Training was as informal as ever, given on the job usually by the mate or an able seaman deckhand, and aiming at enabling the most junior deckhand to take responsibility when the crew needed relief, or in an emergency. A boy who joined the *Nelson* in 1935 made tea while he learned how to relieve the helmsman, just as his predecessors had done. On the *Greetings* in 1938, he decided to move below decks, and started there afresh as a trimmer.

A long-service deckhand needed very few explicit instructions; he knew exactly what to do and when. One applied to Alexandra for a job in 1946, on his release from the army. By then rated as able seaman, he said he could handle a tug, without actually having done so. When he boarded the *Brocklebank* the master was absent, so he asked the mate for a quick tour of the controls, and by the time the skipper returned, he could make good his claim.

It was possible for tugmen to be promoted out of the boats. A chief engineer could go ashore as superintendent engineer, with responsibility for the design and construction of a whole fleet. R. J. 'Mac' McConnell started as a boy in the *Flying Kestrel*, and rose to be manager of Alexandra's Southampton office on the retirement of Stark.

This was not to say that the age-old pitfalls on the way to the top had somehow disappeared. One retired Alexandra officer still remembers a master of the *Nelson* who, in one alcohol-misted week in the 1920s, succeeded in running his tug aground four times.

Although training on the job was still the way of learning its basics, tugmen began to acquire extra qualifications. One reason was the increasing complexity of the technology they were called upon to use. From 1970 Alexandra made on-shore technical training available to their Liverpool crews, and later in other ports as well. The 'boy' still fetched the tea and cooked the meals, but he was likely to be attending vocational classes after work. The 1970s also saw Alexandra growing into coastal and overseas work, which meant that crews had to go on navigational courses, and most masters needed Board of Trade certificates. The tugman of the 1980s was usually a more highly trained man than his ancestors.

The tugmen's union was much more powerful in the 1980s than it had been in the 1920s; but one thing had not changed: the tugmen were as constitutionally at odds with their representatives as ever they had been. The T&GWU had never given its tugmen a single, national voice - they remained organized by ports, with the weakness in negotiation that this entailed. In February 1982 the tugmen created their own United Kingdom Tugboatmen's Association, which soon claimed 2,000 members. They remained members of the T&GWU, which continued to negotiate pay and conditions. The new body was concerned with job protection and standards - a more fundamental and controversial issue, for it involved the use of tugs. The Association was concerned also with safety standards; but its main preoccupations were that British tugs and crews should be employed, and that new, invasive tug operators in a port, with their associated dangers of cheap labour and rate cutting, should be discouraged. This more radical and far-sighted view of the tugman's place in the industry was expected to create new tensions between crews and owners. Tug operators had become accustomed to a workforce concerned almost exclusively with pay, and did not take kindly to what they regarded as illegitimate interference in the way they ran their businesses.

In the mid-1980s, two other traditions were weakening in British ports. There were still many long-service men about - including one Alexandra manager with fifty years under his belt in 1987 - but with a school leaving age of sixteen and compulsory retirement at sixty-five, the astronomical tallies of previous generations were no longer possible. George Robinson of Tees Towing earned the distinction of having a tug named after him while he was still alive and working for them; which was only right and proper, since when he finally retired in 1947, he had been sixty years in tugs. In 1953, W. H. J. Alexander's crews included no fewer than fourteen men with fifty years' service or more. But such very old hands as these were dying out, and not being replaced. Now that more skills were needed of captains, boys stayed at school longer, and commands were fewer, tugmen did not usually get their own tugs until they were about forty. Ernest Watson, master of the even older paddle tug *Lingdale,* had been a skipper for forty-five years when he retired in 1953, having obtained his first command at the age of twenty-four. In the 1980s, few, if any, had their own boats so young.

Company takeovers, together with more mobile tugs, had eroded the old regional traditions once signalled by the tribal divisions of tugmen by port - 'Tilbury folk' - or by company - 'Watkins folk'. Reduced opportunities for promotion at the top of the ladder, and smaller, more highly trained crews that meant fewer openings for the unskilled at the bottom, probably played a part in the decline of the family tradition of tug service. Tug families such as the Alexanders and Russells were harder to find. One of Alexandra's retired masters had had a father, grandfather, and great-grandfather in the tugs. The son and great-nephew of another had been tugmen, and his son still was. But such continuity was increasingly unusual. In the 1980s there was little room for sentiment when a tugman's family decided on careers; and towing now seemed less attractive.

At least the tugman's life was now more comfortable than his father's had been. In a modern tug he had a cabin to himself, and refrigerators and freezers allowed almost as much choice of food and drink at sea as on land. Gone were the days when a tugman used to cook three courses in one pan over the single plate of a coal stove, heating a tin of rice pudding in the water in which the potatoes were boiling; or when fresh water fouled with salt was made drinkable by stirring in some marmalade. One crew had survived thus for five days.

13

Flying Osprey

Looking to the Future

In Liverpool, as Alexandra approached their centenary year, there was a radical slimming-down of the port's traditional operations, in line with the needs of the 1980s. The old docks that survived were in a parlous state; tug owners were paying the Mersey Docks and Harbour Company (as it now was) to keep some of them open. Four more Alexandra crews were made redundant in 1979, and at the end of that year the company's Wellington Dock repair and maintenance shops were closed down and the staff dismissed. This saving, combined with increased towage rates, put Alexandra's Liverpool operation into profit in 1980, but 1981 began with a seamen's strike; then a dispute with the dockers, punctuated by short stoppages, rumbled on for five months from May to September. The Docks Company had been saved from receivership by a Government grant of £15 million, but had been told that there was no more money to come. It was trying to make the port more economic by cutting back the workforce; the issue was pay for the remainder. In the end the 3300 dockers were whittled down to 2500. Meanwhile the port had lost 30 per cent of its general cargo traffic. In June the general cargo handling area of the Birkenhead docks was closed, provoking a dockers' strike on that account, and in September there was a strike at Cammell Lairds. Then, in October, the dockers broke the agreement so painfully reached with a week-long strike. Alexandra's Liverpool revenue, again helped by towage rate increases, was however up slightly in November. Similar events characterized 1982. A fresh agreement between the dockers and the Docks Company, negotiated in September, was followed two months later by a strike. Revenue was hit but improved again following the introduction of the dockers' latest working agreement, which encouraged some ship owners to return to the port. These included a container line and a bulk carrier operator, moving up from strike-crippled Southampton.

The first half of 1983 showed poor results from Liverpool towing, so that

'Discovery' on her way to Dundee with 'Watercock' and 'Ionia'.

more tugs and twenty-four men had to go in June, leaving Alexandra with eleven boats on the Mersey, all busy at peak times, showing that the level was about right. More container lines came to Liverpool; and income was up early in 1984. The rise was counterbalanced by a national dock strike in July; but then revenue looked up again. Although the Mersey picture was not cheerful, it was better than it had been. Alternation of good and bad spells, as at Swansea, was more promising than consistent decline.

The Mersey never fulfilled expectations for the new barges of Bulk Cargo Handling Services, and in 1985 when CPC in Manchester began to take grain from Europe instead of North America (another consequence of the Common Market), the barging business ceased and more than forty men lost their jobs. The future of the Manchester Ship Canal was in doubt too, which led Alexandra to take yet another plunge - BCHS bought and amalgamated two road haulage firms. In 1986 the resulting Alexandra Molyneux Haulage business owned forty road tractors and eighty trailers.

The prosperity of Port Talbot, and therefore of its tugs, was now dependent on the health of the South Wales steel industry: it was even more of a single-industry place than Swansea. Periodic falls in ore imports were a natural consequence of the drops in steel exports that hit Swansea. In 1977 only one ore carrier a week had called, but 5.6 million tons of ore was unloaded in 1979 - a figure which became typical - and by 1985 the number of ore carriers into Port Talbot had doubled.

At Southampton, labour troubles caused a 40 per cent drop in Alexandra's revenue for 1981/82. Because trade was slow in returning, a loss was recorded in the first half of 1983. Although rates went up, revenue fell again in 1984-5 with the renewed dock stoppages. By 1986, fewer and bigger ships had driven down the number of tugs needed at Southampton. In the heyday of the passenger liners and beyond, Alexandra had worked seven tugs and Red Funnel nine: now Alexandra ran six and Red Funnel seven. Slimmed-down fleets sought to eliminate their weaknesses. This sometimes left them at full stretch, for as Alexandra had first found at Liverpool in the 1890s, tugs out at sea could not respond readily to sudden demands in port. There, container ships, tankers, and cargo ships of the older sort all continued to expect help at short notice. In the world shipping crisis, there was also work laying ships up. When the mammoth 458,000-ton tanker *Burmah Endeavour* arrived to be laid up in 1983, the job occupied eight Alexandra and Red Funnel tugs. In June 1986, she moved out again, once more making work for most of Southampton's boats.

Although the trend of work for tugs there was down, Alexandra's confidence in the future of Southampton was signalled when in 1986 they took delivery of two additional boats for the port, to match two recently arrived Red Funnel tugs. These were the large, German-built, second-hand Schottel tugs *Flying Kestrel* and *Flying Osprey* (a new name for the company).

As part of the bicentenary celebrations of Mutiny on the 'Bounty' in 1989, Alexandra-Molyneux Haulage Ltd was responsible for delivering 'Elizabeth Bligh', the replica of Bounty's launch, to the National Maritime Museum, Greenwich, for its send-off and then delivered it to Le Havre.

Bulk Carrier 'Irish Spruce', 72,250 d.w.t. in Port Talbot, this photograph shows clearly the use of the gog rope.

Ups and downs had been the theme of recent Thames business, as at Liverpool and in South Wales. The 'downs' were the result of strikes driving away business, exacerbating the general trend towards fewer, bigger vessels needing fewer tugs. There was an eleven-day stoppage at the end of March 1982, and another major strike in July 1984. For one reason and another, the amount of freight passing through London fell in 1981 to 7 million tons. For Alexandra, 1977 and 1981 showed good profits; while the first half of 1983 saw a loss, owing to a two-month dock strike at Tilbury. The result was twenty-five redundancies on the Gravesend tugs. There were good years, too, as when Alexandra were helping to instal the Thames Barrier. Stretching from bank to bank at Woolwich, it was designed to banish for ever the danger of a flooded London. Work started in 1981. Alexandra tugs towed six flood gates - the largest weighed 1600 tons and was nearly 70 yds long - down from Middlesbrough on the 300 ft by 90 ft construction barges of ITM Offshore Ltd. Each of the Barrier's six concrete sills, one for each gate, was manoeuvred into position by four Alexandra tugs, and one or two from other fleets.

In 1981 optimism was the keynote on the Thames, as elsewhere. The twenty-five boats taken over from London Tugs had been trimmed back to a much healthier seventeen. A new Voith-Schneider tug, the *Sun Thames*, joined the fleet, and another was on order - clear signs of confidence. Alexandra's fleet worked everywhere between the Pool of London and the mouth of the Medway. The *Sun Swale* was stationed as far afield as Ramsgate, at a new cross-Channel ferry terminal with five RoRo (roll on, roll off) vehicle berths. The ULCCs and other big bulk carriers were seen no further upstream than their terminals on the seaward side of Tilbury. There, three years earlier, a second container terminal had been added, enabling Tilbury to handle more than three-quarters of the Australian and New Zealand trade. The completion of the M25 orbital motorway around London boosted the container traffic at Gravesend, now that trucks had easier access. To Tilbury docks came container ships large and small, and the older general cargo ships still run by Third World countries. As on all rivers, business came in peaks and troughs, dictated by the tides - a pattern that had become more pronounced with the coming of bigger ships, all of which came at the same time, trying to enter or leave a berth at the highest point of the tide so as to have as much water as possible under their keels.

Above the Thames Barrier there was comparatively little traffic, with the upstream docks gone. But the surviving wharves took in smaller container ships and general cargo vessels that wanted to come further up-river, and the timber ships still came like the onkers of old. So did cruise liners and visiting warships, up to the Pool and even through Tower Bridge. In 1984 the frigate HMS *Jupiter* sailed as far as London Bridge, the next one upstream. Indeed, at one moment she seemed to want to go beyond the bridge; trying to turn against a flood tide without the help of the attendant Alexandra tugs, she collided with it.

At Felixstowe, Alexandra rose on the back of the port's boom. By 1980 it was handling nearly 400,000 containers a year, which put it in first place among Britain's container ports, ahead of London and Southampton, and this was before spring 1981, when two more container terminals - Walton and Dooley - were opened. In 1981 and 1982 container ship lines were moving in from London, Liverpool and Southampton. The quantity of cargo passing in and out rose from 4.2 million tons at the end of 1975 to nearly 9 million tons in 1984. In that year 4400 ships totalling 33.1 million tons called. Government money was given to help develop a fourth terminal - Trinity - which opened for business in 1985. This multiplicity of terminals all lay within a compact area, which meant that peak time traffic could be handled comfortably in comparison with ports where berths were more widely spread. In turn, this gave the tugs full employment: less time was spent in getting to jobs, which in 1984 averaged two and half per tug per day.

Harwich, across the estuary, was attracting more ships as well, after a programme of dredging and providing deep water berths was completed in 1981. Alexandra responded by sending more tugs. In 1981 the number went up to four (*Alfred, Sun London, Sauria, Egerton*), and in 1983 to five. Three years later the new tractor tug *Brightwell,* with azimuth propellers and 40 tons bollard pull, built at Great Yarmouth, joined the Felixstowe fleet. As well as working at Felixstowe and Harwich, the boats might from time to time be stationed further off, as at Great Yarmouth. Felixstowe towing had provided 5 per cent of Alexandra's total revenue in 1975, after a year in new hands; and in succeeding years the proportion grew.

After years of fluctuating fortunes Gibraltar, too, offered a pleasing prospect with the transfer of the Royal Naval Dockyard to civilian management. In anticipation of the event, Alexandra's strength there was doubled in 1984 to two tugs, the *Sun XXI* and *Deben*.

Under the sea around Britain, new treasures - and potential new sources of income for tugs - were being discovered. In 1981 a gas exploration platform was at work in Morecambe Bay on the Lancashire coast, on Alexandra's doorstep. Built at Cammell Laird, the *Apollo II*'s jacket had been launched with the help of a new *Bramley Moore,* and had been towed out to its drilling site with the aid of three Alexandra tugs. Two of their boats towed out another exploration platform. When production began two years later, ITM Offshore bid for the towage of production jackets, platform construction, and furnishing of supply vessels, but were underbid by the Dutch, who got the contract. Elsewhere ITM did better. Their barges moved two 2000-ton platform sections on the Tyne in 1981, and in 1982, after the Magnus field came on stream, they carried prefabricated harbour sections to Sullum Voe. At this time ITM had four construction barges, including the *ITM Mariner,*

Another shot of 'Irish Spruce' in charge of Alexandra tugs.

Alexandra tugs returning H.M.S. 'Belfast' through Tower Bridge to her permanent berth in the Pool of London. 'Belfast', owned by the Imperial War Museum, is a major tourist attraction (The Press Association, London).

a crane barge capable of lifting 200 tons, and equipped with workshops, a hospital, a helicopter pad, and living accommodation for more than ninety men. By this time Alexandra had a 10 per cent share of ITM. This grew to 20 per cent in 1985. Alexandra were continuing a policy dating back to their purchase of John Cran's shares in 1917, of taking an interest in businesses whose fortunes were tied closely to their own. In 1985, too, ITM ordered a crane barge with a 4000-ton lift - the *ITM Challenger,* which cost £60 million.

Offshore work was extended to Ireland, where in 1979 and 1980 a new port was being built at Foynes, on the river Shannon. An Alexandra tug was on station there for several years.

New and more powerful tugs were constantly added to Alexandra's fleet. In 1978 came the *Indomitable,* a Dunston-built boat with 55 tons of bollard pull and 3500 hp. She and her twin the *Formidable,* delivered in 1979, had Kort rudders with variable pitch propellers. The last boat Dunston built for Alexandra in these years was the *Collingwood* in 1981, a Voith-Schneider harbour tug with 33 tons of bollard pull. A new *Canada,* a Voith-Schneider tug delivered in 1980, was the first boat from McTay Marine of Bromborough on the Mersey. The *Sun Thames* and *Ganges,* harbour tugs that came into service in 1982, were also McTay-built. Later, in 1984 and 1985 respectively, McTay were responsible for the *Sun Anglia* - a sister ship of the *Sun Thames,* with 38 tons of bollard pull - and the *Bramley Moore.* As their names suggested, the tugs prefixed *Sun* were intended for the Thames fleet, as was the *Sun Swale,* a second-hand addition of 1981. The *Ganges,* named after the naval training school at Shotley near Felixstowe, was posted to that port. Another harbour tug, the *Hendon,* had been bought from France Fenwick in 1978, virtually new as it had never been used by them.

The *Formidable* and *Indomitable* were intended to handle the biggest ships afloat, and also construction barges and offshore drilling jackets. But if Alexandra were to win their full share of tug work in the intensely competitive 1980s, they would still need more powerful, seaworthy, and versatile craft, with specialized capabilities, suitable for any offshore construction and service work, anywhere in the world. Therefore, while Alexandra increased the strength of their fleet of harbour tugs, they also introduced a type that was new to them - the supply and anchor-handling boat. They did not build new, for foreign shipyards had been building these craft since the 1970s, and Alexandra wanted boats both quickly and cheaply.

The decision to create the new fleet was taken in 1981. The *Redoubtable,* bought in January 1982, was its first fruit. In power, she was in a different league from her predecessors, disposing of 70 tons of bollard pull and 5000 hp. She had twin Kort rudders with variable pitch propellers, and bow thrusters as well. She and her sister ship the *Implacable,* acquired in 1983, were bought from France. The 495-ton *Invincible,* an anchor-handling tug supply vessel, built in and bought from Norway, also joined in 1983. She was the most powerful of the group, with 8160 hp and 100 tons of bollard pull. The fourth of this class to be added was another French purchase, the *Indefatigable* of 1984, which was identical to the first two. A halt was then called, for there was by now a world surplus of anchor-handling tugs. All the French boats had been built in Yugoslavia in 1975; the *Invincible* was new, built for someone else and bought off the stocks. In January 1984 the *Implacable*, together with the *Canada* and *Collingwood*, towed the jacket of the exploration platform *Sovereign Explorer* from Cammell Laird at Birkenhead to Holyhead, from which point the *Implacable* and *Invincible* towed it to Scotland. A year later the *Formidable* towed another jacket, built by ITM Offshore for a gas field, out of the Tees. By that time the *Invincible* was based at Aberdeen, earning £3000 a day on a yearly contract servicing the North Sea oil fields.

Alexandra's offshore operations were not normally tied to any particular port; and to promote separateness of their identity from traditional harbour towing, in 1980 they became based in new offices in the City of London, at the heart of the nation's trading and shipping business. In 1984 the Alexandra Marine Transportation Company ordered from Portugal a barge with a capacity of 7000 tons, for offshore work: the *AMT Traveller* to complement *AMT Voyager* of similar size.

Now that the biggest tugs, especially, no longer worked at any particular port, their naming was deliberately less parochial. In 1986 the *Indefatigable* was at first stationed at Swansea, then moved to Gravesend. The *Implacable*, too, was a Gravesend boat. But they did not do their main work there: to employ them in moving ships would have been to use a sledgehammer to crack a nut. These places were principally their home bases, from which they ranged the world. With the new armoury at their disposal, Alexandra could bid for work anywhere on earth. The older *Indomitable* and *Formidable*, based on Liverpool and Gravesend respectively, were classed with the big new boats. In 1982 the *Indomitable* was on charter in West Africa, servicing offshore oil work. By 1986 the *Redoubtable* had spent four years almost continuously on the West African coast, working with the Howard Doris crane barge *Tog Mor* off the shores of several countries, constructing, supplying, and moving oil platforms. From 1984 the *Indefatigable* spent time there as well; and undertook a tow from West Africa to Marseilles. During their long postings abroad, the big boats were able to keep in touch with their home bases 5000 miles away thanks to microwave, or super-high frequency, radio transmissions via communications satellite.

In 1985 the *Indomitable* got one of the most unusual jobs done by a tug. She fought her way 600 miles out into the North Atlantic through a full-blooded gale, and there kept station for two and a half days, rolling in a lumpy sea, as she waited to refuel Richard Branson's 60 ft catamaran *Virgin Atlantic Challenger*, which was making an attempt on the Atlantic passenger

'Sun Swale', a second-hand addition to the Thames fleet stationed at Ramsgate. 'Sun' is a traditional prefix for tugs on the Thames.

Fine view of 'Indefatigable' arriving at Swansea.

'Wellington' and 'Alfred' assisting T.&J. Harrison's 'Specialist' (Keith G. Medley, Wallasey).

'Ganges', still on the scene after saving 6 men from the 'European Gateway' on a freezing December night in 1982.

ship speed record held for thirty-three years by the *United States*. The *Indomitable* did her work quickly and efficiently; sadly, the *Virgin Atlantic Challenger* broke up and sank when within the grasp of the record.

Some of the smaller tugs, too, won overseas work - for instance, the *Wellington* and *Waterloo* were towing at the port of Arzew in Algeria in 1981. Though only six years old, they were sold in 1983. A replacement *Wellington* was bought second-hand a couple of years later. The Alexandra tradition of naming boats after Liverpool docks had persisted, fuelling - incidentally - an equally traditional French paranoia about England and English tugs: as late as the 1970s, a French naval liaison officer at Liverpool took grave exception to being sent *Nelson* and *Trafalgar* to move one of his warships. The crew of one tug spent six weeks in Cherbourg - an ancient enemy of tugs from across the Channel - with all its bars closed against them.

The last three of the former Cock boats were sold to new owners in Greece. They sent them around the coast, one old harbour tug towing the other two, through the autumn gales of 1981. They got as far as St Bride's Bay in Pembrokeshire, where three sad little wrecks were cast ashore, looking in their photograph for all the world like children's discarded toys. It was an unsuitable end for a proud line.

The new tugs and barges represented a huge investment. The harbour tugs of 1977 cost £1 million each. The bigger *Indomitable* cost £1½ million in 1978. Yet only two years later a mere harbour tug was priced at £1.4 million; and by 1985 inflation had made the figure £2.3 million. Even an ordinary construction barge, such as the *ITM Voyager* built in 1981, cost £800,000. And bought-in though she was, the *Invincible* set Alexandra back £3.6 million. Between 1979 and 1985, keeping the fleet up to the standard needed to stay in the game called for bank loans totalling nearly £13 million, which had to be serviced and repaid out of revenue.

Alexandra could never afford to relax. In September 1981, Rea went into coastal towing - and specifically, into the Morecambe Bay gas field. Two years later, they absorbed yet another old-established West Country tug operator, C. J. King of Avonmouth. Cory, like Alexandra, won contracts in West Africa, at Luanda and Cabinda in Angola, where they sent two new Z-Peller tugs in 1982, and where they remained for three years. And after the *Implacable* was lost on her way to the Falklands, United Towing stepped in at once with their *Euroman*, in her place. At that time United already had two tugs in the Falklands, so Alexandra had pulled off quite a coup in securing the *Implacable* charter in the first place.

Yet, escalating costs and unrelenting competition notwithstanding, Alexandra came up their hundredth year in fighting condition. The pre-tax group profit for 1979 had been £2.8 million, even though revenue had only been 'ticking over'; and for the first half of 1985 alone, the figure was £1.7 million. The full year produced 'an excellent result'. Between 1979 and 1984 revenue went up from £15.8 million to more than £28 million. And although the glamorous side of the business - the offshore oil and foreign-going work - hogged the limelight, most of the company's profits came from harbour towing. Ships still had to have tugs. This was, perhaps, the most encouraging portent of all.

Euroair Transport grew far beyond the servicing of the needs of Alexandra's towing business. In 1983 the company bought Vickers Viscount airliners, signalling its expansion into internal scheduled services. At the end of 1985 Alexandra sold a half-share of Euroair to Business Air Centre Ltd, a worldwide aircraft charter, management, brokerage, and leasing operation that further expanded Euroair's horizons. By 1986 the Euroair fleet consisted of four Hawker Siddeley 748, three Dart Herald, and five Brazilian-built Embraer Bandeirante airliners, all operating scheduled passenger and freight services; while a Piper Aztec still shuttled Alexandra's crews and managers about. Euroair was in charge of Alastair Mack, son of Graham, who had joined Alexandra in 1975 after experience with Crosthwaite on the Tees, and was moved up to the main board ten years later. In 1982, Dick Braithwaite was appointed managing director of AMT Ltd.

Alex Bicket had died in March 1981, at the great age of ninety-five. He had been a small boy at the time of Alexandra's first major crisis, had nursed the company through its second in the 1930s, and had lived long enough to see it weather the third. In essentials, the industry had not changed in the three generations of towing that Alex Bicket had seen. It remained a business in which small units were more typical than Alexandra's fleet of around fifty tugs - surviving, somehow, all the hazards of insignificance. In the mid-1980s forty or so major owners employed perhaps 3000 people and worked 220 tugs - a drop of only fifty, in thirty years of shipping decline. They had found new work to do. If one-man operations were taken into account - and there were still plenty of them - the number of operators rose to well over 100 and the number of tugs to 400 or more. If the towing industry was sick, as some suggested, it was sick of the same diseases it had suffered for a century and a half, and had triumphantly survived.

Invincible

Sources

Newspapers and Periodicals

Boston Log, The, Dec 1971
Economic History Review Vol 9 2nd series 1956, pp. 74-88
Economist, The
Empire Illustrated, The, supplement, Ninety Years of Towing, 1923
Fairplay
Financial Times, The
Glasgow Herald, The
Gravesend and Dartford Reporter, The
Hampshire Magazine, July 1984
Herald of Wales, The
Illustrated London News
Industries and Iron, 18 September 1896
Journal of Commerce and Shipping Telegraph, The
Lekko
Lighthouse, The, 7 February 1891
Liverpool Courier, The
Liverpool Mercury, The
Liverpool Nautical Research Society, Transactions
Lloyd's List
Mariner and Engineering Record, The
Mariners' Mirror, The
Mona's Herald, 25 July 1860
Pall Mall Magazine, The, November 1894, vol. 4, no. 19
PLA Monthly
Scotsman, The, 2 January 1884
Sea Breezes
Shipping Gazette, The
Southampton Port News

Southern Evening Echo, The
South Wales Evening Post, The
Teesside Journal of Commerce, The, November 1953
Times, The
Western Mail, The

Books

Anderson, Roy, *White Star*, Prescot, 1964
Armitage, W.F. *Tugs and Towing Operations*, London, 1962
Batey, Frank, *Tugs, their Work and Development*, London
Bowen, Frank C. *A Hundred Years of Towage 1833-1933*, Gravesend, 1933
Butlin, M., and Joll, E., *The Paintings of J.M.W. Turner*, London, 1977
Cable, Boyd, *A Hundred Year History of the P&O*, London, 1937
Caldwell, A., *Steam Tug Design*, London, 1969
Chandler, G., *Liverpool Shipping, A Short History*, London, 1960
Chatterton, E. Keble, *The Epic of Dunkirk*, London, 1940
Clapham, J.H., *Economic History of Modern Britain*, Cambridge, 1938
Conrad, Joseph, *The Mirror of the Sea*, 1906
Corlett, Ewan, *The Iron Ship*, Bradford on Avon, 1975
Cornewall-Jones, R.J., *The British Merchant Service*, London, 1969
Course, A.G., *The Merchant Navy, a Social History*, London, 1963
Craig Robin, *Steam Tramps and Cargo Liners 1850-1950*, London, 1980
Duff, Peter, *British Ships and Shipping*, London, 1949
Fayle, C.E., *The War and the Shipping Industry*, London, 1927
Gores, Joseph N., *Marine Salvage*, Newton Abbot, 1972
Hallam, W.B., *Blow Five, A History of The Alexandra Towing Co Ltd*, Liverpool, 1976
Hallam, W.B., *Fifty Years of Mersey Towage 1916-1966*, Liverpool, 1966
Hardy, A.C., and Tyrrell, E., *Shipbuilding*, London, 1964
Harvey, W.S., and Downs-Rose, G., *William Symington*, London, 1980
Hazell, R.F., *Gaselee & Sons Ltd 1879-1954*, Reading, 1954
Heaton, Thomas, *Tugs*, Liverpool, 1923
Hyde, F.E., *Blue Funnel*, Liverpool, 1956
Hyde, F.E., *Liverpool and the Mersey 1700-1970*, Newton Abbot, 1971
Kemp, Peter, *The Oxford Companion to Ships and the Sea*, Oxford, 1976
Kennedy, N.W., *Record of Early British Steam Ships*, Liverpool, 1933
Kirkaldy, A.W., *British Shipping*, London, 1913
Le Gillou, M.A., *A History of the River Tees 1000-1975*, Cleveland, 1978
Lubbock, A.B., *The Blackwell Frigates*, Glasgow, 1950
Lubbock, A.B., *The Western Ocean Packets*, Glasgow, 1925
McKee, Alexander, *The Golden Wreck*, London, 1961
McMurray H. Campbell, *Old Order, New Thing*, Greenwich, 1972
Maginnis, A.J., *The Atlantic Ferry*, London, 1893

Masefield, John, *The Bird of Dawning*, London, 1933
Masefield, John, *Salt Water Ballads*, London, 1902
Masters, David, *When Ships Go Down*, London, 1932
Mersey Docks & Harbour Board, *Business in Great Waters 1858-1958*, Liverpool, 1958
Mountfield, A.S., *Western Gateway: A History of the Mersey Docks and Harbour Board*, Liverpool, 1965
O'Brien, F.T., *Early Solent Steamers*, Glasgow, 1981
O'Connor, G.W., *The First Hundred Years: The Southampton, Isle of Wight and South of England Royal Mail Steam Packet Company 1861-1961*, Southampton, 1962
Orchard, B.G., *Liverpool's Legion of Honour*, Birkenhead, 1893
Owen, D.J., *Origin and Development of Ports in the United Kingdom*, London, 1948
Padfield, Peter, *Beneath the House Flag of the P&O*, London, 1981
Port of Felixstowe Handbook 1985, 1986
Proud, John H., *Sea Horses of the Tees*, Middlesbrough, 1985
Riddell, John F., *Clyde Navigation, A History of the Development of the River Clyde*, Edinburgh, 1979
Rowland, K.T., *Steam at Sea*, Newton Abbot, 1970
Smith, Edgar C., *A Short History of Naval and Marine Engineering*, Cambridge, 1938
Taplin, Eric L. *Liverpool Dockers and Seamen 1870-1890*, Hull, 1974
Thomas, P.N., *British Steam Tugs*, Wolverhampton, 1983
Thornton, R.H., *British Shipping*, Cambridge, 1959
Tyneside Industries, 1889
Williams, Mark, *No Cure, No Pay, The Story of Salvage at Sea*, London, 1978
Wilsher, Peter, *The Pound in your Pocket*, London, 1970
Young, G.M. (ed) *Early Victorian England*, Oxford, 1934

Official Publications

Abstract of Statistics, CSO
Court of Inquiry Concerning Transport Workers' Wages and Conditions of Employment of Dock Labour, Report and Minutes, 1920
Royal Commision on Labour, Report and Minutes, 1892-4

Unpublished Records

Alexander, W.H.J., logs
Alexandra Towing Company, minutes of board meetings and general meetings, annual reports, account books, logs, particulars of tugs (various owners).
Flying Breeze, master's day book, 1914 (manuscript)
History of Rea Ltd., Liverpool (typescript)

Hallam, W.B., *Mersey Tugs*, 1955 (typescript)
Knight, M.T. *A History of J.P. Knight Ltd* (typescript)
Mayor, F.J., *Development in Thames Tugs*, transcript of paper read before Diesel Engine Users' Association, Jan 18 1951 (typescript)
North Isle to The Alexandra Towing Company Southampton, transcript of radio report 24-26 October 1970 (*Pacific Glory*)
Scott, G.R.D., *W.H.J. Alexander Ltd 1883-1953*, 1953 (typescript)
Simmons, Captain William, biographical notes (manuscript)
Sun XV to owners, report 30 May-3 June 1940, manuscript (Dunkirk)
William Watkins Ltd, particulars of tugs.

Index

Notes: Vessels named after people are indexed by Christian name, eg *William Poulsom*.
Company names are listed by surname, eg Cran, John, & Co.

Netherfield Road
Everton Terrace
Netherfield Road South
Everton Village
St. Anthonys Church
St. Martins Church
Great Homer St.
Roscommon St.
St. Augustines Church
Collegiate Inst.
Shaw Street
Soho
Saint Ann Street
Christian Street
Norton Street
Richmond Row
Byrom Street
St. Georges Hall
Railway Entrance
Lime Street
St. Johns Ch.
St. Johns Lane
Whitechapel
Tithebarn Street
Moorfields
Dale Street
Town Hall
Water Street
Chapel St.
St. Nicholas Ch.
Goree
Brunswick St.
James Street
Warehouses
Strand Street
Canning Place
New Custom Hou.
Hanover Street
Sailors Home
East Side Salthouse Dock
Dukes Place
Canning Dock
Georges Dock
Salthouse Dock
Public Baths
Georges Dock Parade
Manchester Dock
Albert Dock Parade
Basin
Dukes Yard & Pier
Birkenhead Hotel